W9-AVG-400

GIS AND GENERALIZATION

Methodology and Practice

Also in the GISDATA SERIES

II ***Spatial analytical perspectives on GIS in the environmental and socio-economic sciences***
edited by M. Fischer, H. Sholten and D. Unwin

III ***Geographic objects with undetermined boundaries***
edited by P. Burrough and A. Frank

Series Editors

I. Masser and F. Salgé

GIS AND GENERALIZATION

Methodology and Practice

Editors

J. C. Müller, J. P. Lagrange and R. Weibel

GISDATA I

Series Editors

I. Masser and F. Salgé

UK Taylor & Francis Ltd, 4 John St, London WC1N 2ET

USA Taylor & Francis Inc., 1900 Frost Road, Suite 101, Bristol, PA 19007

British Library Cataloguing in Publication Data

A catalogue record for this book is available from the British Library

ISBN 0-7484-0318-3 (cased)
0-7484-0319-1 (paper)

Library of Congress Cataloging in Publication Data are available

Cover design by Hybert Design and Type

Typeset by Keyword Typesetting Services Ltd, Wallington, Surrey

Printed in Great Britain by Burgess Science Press, Basingstoke, on paper which has a specified pH value on final paper manufacture of not less than 7.5 and is therefore 'acid free'.

Contents

Series Editors' Preface vii

Editors' Preface ix

Contributors xiii

SECTION I INTRODUCTION 1

1 Generalization: state of the art and issues 3
J. C. Müller, R. Weibel, J.P. Lagrange and F. Salgé

SECTION II GENERIC ISSUES 19

2 GIS-based map compilation and generalization 21
S. Morehouse

3 The need for generalization in a GIS environment 31
E. Spiess

4 Development of computer-assisted generalization on the basis of cartographic model theory 47
D. Grünreich

5 Three essential building blocks for automated generalization 56
R. Weibel

SECTION III OBJECT-ORIENTED AND KNOWLEDGE-BASED MODELLING 71

6 Data and knowledge modelling for generalization 73
A. Ruas and J.P. Lagrange

7 Object-oriented map generalization: modelling and cartographic considerations 91
B.P. Buttenfield

8 Holistic generalization of large-scale cartographic data 106
G.L. Bundy, C.B. Jones and E. Furse

9 The GAP-tree, an approach to 'on-the-fly' map generalization of an area partitioning 120
P. van Oosterom

SECTION IV KNOWLEDGE ACQUISITION 133

10 Potentials and limitations of artificial intelligence techniques applied to generalization 135
S.F. Keller

11 Rule-orientated definition of the small area 'selection' and 'combination' steps of the generalization procedure 148
M. Heisser, G. Vickus and J. Schoppmeyer

12 Knowledge acquisition for cartographic generalization: experimental methods 161
R.B. McMaster

SECTION V DATA QUALITY 181

13 The importance of quantifying the effects of generalization 183
E.M. João

14 The effects of generalization on attribute accuracy in natural resource maps 194
M. Painho

SECTION VI OPERATIONAL ISSUES 207

15 Incremental generalization for multiple representations of geographical objects 209
T. Kilpeläinen and T. Sarjakoski

16 Experiment on formalizing the generalization process 219
D. Lee

17 A hierarchical top-down bottom-up approach to topographic map generalization 235
G.J. Robinson

18 Strategies for ATKIS-related cartographic products 246
G. Vickus

Index 253

Series Editors' Preface

The GISDATA series

Over the last few years there have been many signs that a European GIS community is coming into existence. This is particularly evident in the launch of the first of the European GIS (EGIS) conferences in Amsterdam in April 1990, the publication of the first issue of a GIS journal devoted to European issues (*GIS Europe*) in February 1992, the creation of a multi-purpose European ground-related information network (MEGRIN) in June 1993, and the establishment of a European organization for geographic information (EUROGI) in October 1993. Set in the context of increasing pressures towards greater European integration, these developments can be seen as a clear indication of the need to exploit the potential of a technology that transcends national boundaries to deal with a wide range of social and environmental problems that are also increasingly seen as transcending national boundaries within Europe.

The GISDATA scientific programme is very much part of such developments. Its origins go back to January 1991, when the European Science Foundation funded a small workshop at Davos in Switzerland to explore the need for a European level GIS research programme. Given the tendencies noted above it is not surprising that participants of this workshop felt very strongly that a programme of this kind was urgently needed to overcome the fragmentation of existing research efforts within Europe. They also argued that such a programme should concentrate on fundamental research and it should have a strong technology transfer component to facilitate the exchange of ideas and experience at a crucial stage in the development of an important new research field. Following this meeting a small coordinating group was set up to prepare more detailed proposals for a GIS scientific programme during 1992. A central element of these proposals was a research agenda of priority issues grouped together under the headings of geographic databases, geographic data integration, and social and environmental applications.

The GISDATA scientific programme was launched in January 1993. It is a four-year scientific programme of the Standing Committee of Social Sciences of the European Science Foundation. By the end of the programme more than 300 scientists from 20 European countries will have directly participated in GISDATA activities and many others will have utilized the networks built up as a result of them. Its objectives are:

- to enhance existing national research efforts and promote collaborative ventures over coming European-wide limitations in geographic data integration, database design and social and environmental applications;
- to increase awareness of the political, cultural, organizational, technical and informational barriers to the increased utilization and inter-operability of GIS in Europe;
- to promote the ethical use of integrated information systems, including GIS, which handle socio-economic data by respecting the legal restrictions on data privacy at the national and European levels;

- to facilitate the development of appropriate methodologies for GIS research at the European level;
- to produce output of high scientific value;
- to build up a European network of researchers with particular emphasis on young researchers in the GIS field.

A key feature of the GISDATA programme is the series of specialist meetings that is being organized to discuss each of the issues outlined in the research agenda. The organization of each of these meetings is in the hands of a small task force of leading European experts in the field. The aim of these meetings is to stimulate research networking at the European level on the issues involved and also to produce high quality output in the form of books, special issues of major journals and other materials.

With these considerations in mind, and in collaboration with Taylor & Francis, the GISDATA series has been established to provide a showcase for this work. It will present the products of selected specialist meetings in the form of edited volumes of specially commissioned studies. The basic objective of the GISDATA series is to make the findings of these meetings accessible to as wide an audience as possible to facilitate the development of the GIS field as a whole.

For these reasons the work described in the series is likely to be of considerable importance in the context of the growing European GIS community. However, given that GIS is essentially a global technology most of the issues discussed in these volumes have their counterparts in research in other parts of the world. In fact there is already a strong US dimension to the GISDATA programme as a result of the collaborative links that have been established with the National Center for Geographic Information and Analysis through the National Science Foundation. As a result it is felt that the subject matter contained in these volumes will make a significant contribution to global debates on geographic information systems research.

Ian Masser and François Salgé

Editors' Preface

GIS and Generalization: methodology and practice

This book comes out as the result of a specialist meeting in Compiègne, France, in December 1993. The meeting was sponsored by the European Science Foundation (ESF) Social Science Program through the GISDATA Programme, and initiated by the GISDATA Steering Committee which recognized generalization of geographic information as one of the major unresolved issues in geographic information systems. The chapters of this book represent the contributions of the experts who gathered at the Compiègne meeting. As such, we feel that this compilation of articles represents a major effort in describing the state of the art in scientific generalization. It also provides a perspective on future research priorities, such as model generalization for the manipulation of geographic objects independently from their graphic representations.

This book comes at a time when the prospects for automation in the generalization of geographic information are still uncertain. There is no doubt that part of the generalization process can be automated, as shown by several authors reporting on various experiments and commercial software already implemented. But the extent to which automation will be achieved in the future is still unknown. So far, we have not gone much beyond the implementation of geometric operators which are used automatically to generalize individual geographic features such as street networks, coastlines and population settlements. Therefore, this book offers both a speculative view on ways to implement automated generalization, and practical examples of already implemented or directly implementable generalization software for the purpose of map production.

Generalization is motivated by the need to provide multiple views of geographical data at various scales and levels of resolution. It is a tool which has been in use for many years, particularly in cartography for the production of maps at smaller scales. What is new, however, is the context in which this tool operates. The introduction of geographical information systems (GIS) for planning purposes and automated map production in mapping agencies provides a new *raison d'être* to generalization, namely as a tool to model data to support either spatial analysis or the automated production of maps at multiple scales. Hence, the flavour of this book, which discusses generalization issues from a model building perspective. Generic problems, modelling and knowledge acquisition issues are presented first, before discussing data quality and production issues. A major difficulty for the implementation of automated procedures is the acquisition and the formalization of geographic and cartographic knowledge (essentially intuitive and fuzzy), which has been used for many years in manual generalization. Various views are expressed, some defending a holistic approach to generalization, others favouring a more practical stepwise approach. The opinion, however, that semantic object definition, geographical analysis and knowledge formalization is a necessary prerequisite to resolving graphical conflicts in map representation is strongly represented throughout the book. The concerns about loss of data quality, which potentially occurs through generalization, are also taken into consideration, since generalized information may lead to distorted views of reality.

The book is organized in six parts; Part I Chapter 1 by Müller *et al.* is an introduction to all following parts. It contains the position paper by the members of the organizing Task Force that was distributed to all participants prior to the meeting, requesting them to write their own discussion papers in response to the issues that were addressed in this article.

Part II on generic issues of generalization contains papers that provided the major elements for discussion at the meeting. Chapters 2 by Morehouse and 3 by Spiess are the two keynote presentations that were given at the meeting. The invited speakers were deliberately chosen so that their respective background and approach towards the automation of generalization would differ. Thus, even in these two contributions, some of the tensions, but also design options that are typical to generalization, become apparent. Based on their experience with previous projects, Grünreich (Chapter 4) and Weibel (Chapter 5) discuss the fundamental importance of including the relation between cartographic and model generalization, methods for knowledge acquisition, and quality assessment.

Part III deals with object-oriented methods and knowledge-based modelling. Chapter 6 by Ruas and Lagrange opens the discussion debating the implications that the choice of data representations (i.e. data models) may have on the knowledge that can be represented in the respective systems, and thus the tasks that can be fulfilled. Buttenfield (Chapter 7) stresses the need for appropriate data models to support map generalization, whereas Bundy *et al.* (Chapter 8) and Oosterom (Chapter 9) propose various alternatives to structure the data in order to facilitate automated generalization.

The chapters of Part IV then focus on knowledge acquisition and representation, issues that are surely gaining in importance. Keller in Chapter 10 reviews techniques of artificial intelligence as they are available in computer science, and discusses how they could be made useful in the context of generalization. Chapter 11 by Heisser *et al.* presents a particular study in conjunction with the German ATKIS project. It shows how in a sufficiently defined and constrained situation, knowledge can be deduced from the definitions and constraints and integrated with generic cartographic knowledge. McMaster (Chapter 12) concentrates on procedural knowledge (i.e. generalization operators and their choice and parametrization) and on possibilities of using interactive systems for acquisition of this particular type of knowledge.

Part V deals specifically with data quality issues related to generalization activities. João (Chapter 13) discusses the importance of quantifying and controlling the effects of model and cartographic generalization on GIS map manipulations, and Painho (Chapter 14) presents a case study where the effects of generalization on the attribute accuracy of a small-scale vegetation map are analysed.

Part VI concludes the book with operational and implementation issues. Kilpeläinen and Sarjakoski (Chapter 15) propose an incremental approach, where updates after generalization can be propagated through the entire geographic database. Lee (Chapter 16) shows that commercial software can be successfully developed for semi-automated generalization. This software comes in the form of a tool box where the tools are generalization operators, which can be used to manipulate individual map features. Robinson (Chapter 17) and Vickus (Chapter 18) give their view on generalization strategies, requirements and development needs in two mapping agencies (Ordnance Survey, and Survey and Mapping in North-Rhine Westphalia).

As editors of this book, we take the opportunity to thank all authors for their contribution, as well as the various anonymous reviewers who helped in improving the quality of the articles. Books dealing with GIS and software issues have necessarily a short life. We hope, however, that it provides for the next few years both encouragement and strong impulse to further theoretical and practical research in this important area, and that the concerns expressed in this book will influence the design of future commercial geographic information systems. We hope in particular that it leads the way to generalization solutions which will make it possible to concentrate on the application of generalization in GIS, once the theory has been mastered.

Jean-Claude Müller, Jean-Philippe Lagrange and Robert Weibel

Contributors

Geraint Bundy
Department of Computer Studies, Pryfysgol Morgannwg/University of Glamorgan, Pontypridd, Mid Glamorgan CF37 1DL, UK

Barbara Buttenfield
NCGIA State University of New York, Buffalo, NY 14261, USA

Edmund Furse
Department of Computer Studies, Pryfysgol Morgannwg/University of Glamorgan, Pontypridd, Mid Glamorgan CF37 1DL, UK

Dietmar Grünreich
Institute of Cartography, University of Hannover, Appelstrasse 9a, 30167 Hannover 1, Germany

Michael Heisser
Institut für Kartographie und Topographie, Universität Bonn, Meckenheimer Allee 172, 53115 Bonn, Germany

Elsa Maria João
Department of Geography, London School of Economics, Houghton Street, London WC2A 2AE, UK

Chris B. Jones
Department of Computer Studies, Pryfysgol Morgannwg/University of Glamorgan, Pontypridd, Mid Glamorgan CF37 1DL, UK

Stefan Keller
Unisys (Schweiz) AG, Zuercherstrasse 59–61, CH-8800 Thalwil, Switzerland

Tina Kilpeläinen
Department of Cartography and Geoinformatics, Finnish Geodetic Institute, Imalankatu 1A, SF-00240 Helsinki, Finland

Jean-Philippe Lagrange
IGN, 2 Avenue Pasteur, 94160 Saint Mandé, France

Dan Lee
Mapping Science Division, Intergraph Corp., 2051 Mercator Drive, Reston, VA 22091, USA

Robert B. McMaster
Department of Geography, University of Minnesota, 414 Social Sciences Building, Minneapolis, MN 55455, USA

Scott Morehouse
Environmental Systems Research Institute, 380 New York Street, Redlands, CA 92373, USA

Jean-Claude Müller
Geographisches Institut, Ruhr Universitaet Bochum, Universitaetsstrasse 150, Gebaeude NA, D-44780 Bochum, Germany

P.J.M. van Oosterom
TNO, Physics and Electronics Laboratory, PO 96864, 2509 JG The Hague, The Netherlands

Marco Painho
The Higher Institute for Statistics and Information Management (ISEGI), New University of Lisbon (UNL), Travessa Estêvão Pinto, Campolide, 1000 Lisboa, Portugal

Gary J. Robinson
NERC Unit for Thematic Information Systems, Department of Geography, University of Reading, Reading RG6 2AB, UK

Anne Ruas
Service de la Recherche-IGN, 2 Avenue Pasteur, 94160 Saint Mandé, France

François Salgé
IGN, 2 Avenue Pasteur, 94160 Saint Mandé, France

Tapani Sarjkoski
Department of Cartography and Geoinformatics, Finnish Geodetic Institute, Imalankata 1A, SF-00240 Helsinki, Finland

Johannes Schoppmeyer
Institut für Kartographie und Topographie, Universität Bonn, Meckenheimer Allee 172, 53115 Bonn, Germany

Ernst Spiess
Institut für Kartographie, ETH Hönggerberg, CH-8093 Zurich, Switzerland

Georg Vickus
Landesvermessungsamt Nordrhein-Westfalen, Muffendorfer Strasse 19–21, 53177 Bonn, North-Rhine-Westphalia, Germany

Robert Weibel
Department of Geography, University of Zurich, Winterthurerstrasse 190, CH-8057 Zurich, Switzerland

The **European Science Foundation** is an association of its 55 member research councils, academies, and institutions devoted to basic scientific research in 20 countries. The ESF assists its Member Organizations in two main ways: by bringing scientists together in its Scientific Programmes, Networks and European Research Conferences, to work on topics of common concern; and through the joint study of issues of strategic importance in European science policy.

The scientific work sponsored by ESF includes basic research in the natural and technical sciences, the medical and biosciences, and the humanities and social sciences.

The ESF maintains close relations with other scientific institutions within and outside Europe. By its activities, ESF adds value by cooperation and coordination across national frontiers and endeavours, offers experts scientific advice on strategic issues, and provides the European forum for fundamental science.

This volume is the first of a new series arising from the work of the ESF Scientific Programme on Geographic Information Systems: Data Integration and Database Design (GISDATA). This four-year programme was launched in January 1993 and through its activities has stimulated a number of successful collaborations among GIS researchers across Europe.

Further information on the ESF activities in general can be obtained from:
European Science Foundation
1 quai Lezay Marnesia
67080 Strasbourg Cedex
tel: +33 88 76 71 00
fax: +33 88 37 05 32

EUROPEAN SCIENCE FOUNDATION

This series arises from the work of the ESF Scientific Programme on Geographic Information Systems: Data Integration and Database Design (GISDATA). The Scientific Steering Committee of GISDATA includes:

Dr Geir-Harald Strand
Norwegian Institute of Land Inventory
Box 115, N-1430 Ås
Norway

Dr Antonio Susanna
ENEA DISP-ARA
Via Vitaliano Brancati 48
00144 Roma
Italy

SECTION I

Introduction

1

Generalization: state of the art and issues

J.C. Müller,[a] R. Weibel,[b] J.P. Lagrange[c] and F. Salgé[c]

[a]*Geographisches Institut, Ruhr Universität Bochum, Universitätstraße 150, 44780 Bochum, Germany*
[b]*Department of Geography, University of Zurich, Winterthurerstraße 190, 8057 Zurich, Switzerland*
[c]*IGN, 2 Avenue Pasteur, 94160-Saint Mandé, France*

1.1 Introduction

This chapter intends to provoke discussions and reactions on a number of items relevant to GIS data visualization at multiple levels of scale (the ratio between the size of an object on the map and its real size on the ground) and resolution (the smallest object which can be represented on the map).

From a user point of view, visualization is the window of GIS and is essential for visual data exploration, interpretation and communication. Geographical processes are scale dependent and numerous applications in climate, water resources, agriculture, forestry, transportation, land and urban planning require changing degrees of detail and generalization when analysis and communication occur at the local or more global levels. Hence, there is a need for the modelling of geographic information at different levels of abstraction. Ideally, one should be able to view and analyse data at the level where geographical variance is maximized (Tobler and Mollering, 1972; Woodcock and Strahler, 1987) or where spatial processes are best understood.

From a data production point of view, the management and maintenance of spatial data are constrained by the requirements for accuracy, i.e. 'relationship between a measurement and the reality which it purports to represent' (Goodchild, 1991); precision, i.e. 'degree of detail in the reporting of a measurement' (Goodchild, 1991); and quality control. Requirements for the flexibility afforded by multiple scale production and update operations complicate the issues of accuracy, consistency and integrity.

The question, therefore, is not whether geographic information (in digital or analogue forms) should be made available at multiple levels of abstraction, but how it should be made available.

1.2 Pro- or contra-generalization?

Some authors argue that generalization — in the cartographic sense — is not a prerequisite to the delivery of geographic information at multiple levels of scale and resolution.

The ability of current GISs to zoom in and out of a given area, to break down a single multi-thematic layer into a series of mono-thematic layers, and concurrently to produce multiple windows of the corresponding zooming and layering operations, explains perhaps the historical lack of interest of the GIS community in cartographic generalization. Most GIS commercial firms have denied or ignored the cartographic generalization issue.

Other authors admit that generalization would be a useful tool in the GIS tool-kit but argue that automated generalization is either an 'NP-complete' problem (i.e. a computational solution cannot be devised) or the practical and economic benefits of a solution are dubious. The first view is strong among conventional cartographers. The latter view is shared by many national mapping agencies (NMAs) which store multiple scale versions of manually generalized data. In smaller countries like The Netherlands, where the size of the map series is rather limited, the inconvenience of storage overheads and duplication in updating efforts is perceived as a lesser evil compared to the potential processing cost of an automated generalization solution which has not yet arrived.

For larger countries like France, NMAs are still forced to store multiple scale versions, for a number of reasons: there is no production tool for generalization (on the market place) able to derive the required datasets; there is no tool to propagate updates through a series of derived datasets; the processes of regenerating datasets are expensive and require a long time (hence it is not profitable to carry those processes in an industrial context except once, which explains why the various datasets are maintained more or less separately). Finally, the smaller the scale, the shorter the update cycle. Hence, if a NMA wants to maintain only one scale version then it has to update it frequently, with the higher geometric accuracy, in order to respond to the update needs for all other smaller scale versions. This is a dilemma faced by NMAs which goes against the idea of one single database (the expression 'scaleless or scale-free database', which may lead to confusion, is purposefully avoided here; it seems that in the case of data coming from surveys or photogrammetry, it would be more appropriate to speak of precision, accuracy and resolution, not scale, since the notion of scale is meaningless in the absence of a mapping relation).

As a result, most research efforts to resolve the problem of automated generalization, whether in the context of GIS or for the production of paper maps, have been confined to academia. Some academics have argued that the storage of a finite series of multiple scale cartographic databases provides major impediments from both a scientific and management point of view. From a scientific view point because 'what if' and 'if–then' scenarios in GIS require the possibility of navigating dynamically and continuously from any scale to any other scale automatically. From a management point of view because one cannot afford the duplication of efforts that occur in map series updating as well as the inconsistencies which may arise through this process. Solutions to these impediments require moving beyond the paradigm of traditional paper map series, without sacrificing support for the production of paper maps. That is to say, the generation of digital products can no longer be driven by paper map production, as the needs for spatial data have become much broader and complex. Generalization facilities must be provided by GISs to support the use of geographic information at multiple scales for multiple purposes and tasks. Note that besides GIS electronic displays, paper maps will continue to exist; paper maps are permanent, transportable documents, and they also offer a far better accuracy and can represent more data than the screen of a CRT. Furthermore, maps, as visual communication means, are still the easiest and quickest to read media for communicating geographical information to the reader.

Some mapping agencies and commercial firms are now investing resources to implement automated or semi-automated generalization facilities. Apart from the effect of individual leadership, this could mean that there is a growing belief among professionals that generalization could become an operational tool for the production of geographic information in the years to come. Professionals have also come to realize that the full potential of GIS can only be exploited if functions for automated generalization are available.

Assuming that the answer to the previous hypothesis is true — that is generalization in the context of GIS and automated map production is both desirable and feasible — we are now faced with the theoretical and practical issues of building systems for automated generalization. Some authors have used the term 'generalization machines' (e.g. João *et al.*, 1993) to describe such systems; the term has a strong mechanistic connotation, however, and we wonder whether we can use it in this context.

1.3 Generalization yes, but what are the issues?

We need to distinguish between the issues that are brought about by graphical representation from those which arise from modelling at different levels of spatial and semantic resolution. Generalization may be viewed as an interpretation process which leads to a higher level view of some phenomena — looking at them 'at a smaller scale'. This paradigm is always the first used in any generalization activity, whether spatial or statistical. Second, generalization can be viewed as a series of transformations in some graphic representation of spatial information, intended to improve data legibility and understanding, and performed with respect to the interpretation which defines the end-product. These two categories have motivated research mainly in two areas: model-oriented generalization, with focus on the first stage above-mentioned, and cartographic generalization, which deals with graphic representation.

Issues relevant to graphical representations are well known to conventional cartographers. In geographical circles, people usually think of generalization as part of cartographic compilation whose purpose is to resolve legibility problems. An operation such as feature displacement is typically cartographic. Should we go beyond this and consider generalization in contexts which are not necessarily representational? The distinction, for example, between cartographic and statistically controlled generalizations was made before (Brassel and Weibel, 1988). Modelling reliability on statistical surfaces by polygonal filtering (Herzog, 1989) is not necessarily directed towards visualization but helps to understand data by providing higher levels of abstraction. In this case, the motivation as well as the solutions to bring about the necessary transformations are not the same as for cartographic generalization. Generalization in the sense of modelling is a requirement for spatial analysis and the tools (e.g. spatial districting and aggregation of spatial enumeration units, image classification, trend surface analysis, surface filtering, and kriging) have already been developed (see, in particular, Tobler, 1966; Tobler and Moellering, 1972). Do we need to consider this category of generalization in our research agenda? Is it relevant to the producers of geoinformation? Is there a need for future research? Or should we close the book on statistically oriented generalization instead?

The first part of this position paper deals with data abstraction, i.e. a reduction of spatial as well as semantic resolution, whether motivated by data analysis or cartographic representation. We will coin this kind of activity under the general term 'model-oriented generalization'.

1.4 Model-oriented generalization

The difference between the model view and the cartographic view of generalization is the possibility for database manipulation in the former case, independent of cartographic representation. Spatial objects may need to have multiple digital representations in which internal representations (models) should be distinguished from visualization (cartographic) representations. One reason for generalization at the modelling level is to facilitate data access in GIS. This need becomes urgent in view of the design of GISs in which the user interacts with the geo-objects without knowledge of their internal representation. Also, model generalization may be driven by analytical queries (Where are the trends?, What is the spatial average?, Where are the new classes to appear at this level of variance?, etc.) whereas cartographic generalization is mainly driven by communication requirements (legibility, graphical clarity, and understandability). But the two types are not independent, and one (model-oriented) can be a precursor to the other (graphics-oriented). The question is how much and what kind of model-oriented generalization support is required for the accomplishment of routine tasks in cartographic generalization?

In model-oriented generalization, methods are currently being developed to support insertions, deletions, updates and geometric queries at an arbitrary location for an arbitrary scale (Becker *et al.*, 1991). A generalization index may be applied to point data which, in turn, defines their priority for access or rescaling operations. Storage structures for seamless, 'scaleless' geographic databases have also been proposed (Oosterom, 1989). Hierarchical data structures, including quad trees and strip trees, are often used to subdivide and merge data for generalization purposes (Jones and Abraham, 1986). The working hypothesis put forward by database experts is that spatial proximity information must be implicitly available in order to favour access to local information and neighbourliness relationships.

The basic categories of space found in the GIS literature, namely metric, topological, and structural categories, can be used to describe various levels of abstraction for spatial objects. The metric space describes distance relations and constitutes the lowest level of abstraction. The topological space, instead, deals with the existence of spatial relations between points in space. The highest level of abstraction is reached through the structural space which only deals with entities and relations (Sowa, 1984). Abstraction of a road network using hypergraphs and graph theoretic concepts is an example of structural representation (Titeux, 1989; Salge *et al.*, 1990). The question is whether we can invent protocols to propagate changes (say through updating) from one level of abstraction to all others. This would go a long way towards detecting inconsistencies between representations.

Other models for data abstraction and data structuring are also available, but are still in the laboratories. For instance, what are the potentials of abstraction mechanisms known from semantic modelling (Smith and Smith, 1977; Hull and King, 1987), including classification, generalization, aggregation and association, in formalizing relations between spatial objects? There has been much excitement about the introduction of object-oriented programming in GIS. Apart from the confusion surrounding the idea of object orientation, most GIS vendors use the concept for advertising purposes. The object-oriented environment, where procedures (methods) are bound to the object, objects communicate with each other and inherit attributes and methods from others, seems to offer great potentials for implementing generalization procedures. The concept of 'delegation between objects', in particular, could be used to perform updates concurrently across all map-scale layers in the database. As with semantic modelling, the proposed models are attractive but have no proven records yet in the field of generalization.

Temporal abstraction is another type of data modelling which expresses changes occurring in spatial objects (and their attributes) at different intervals of time (Langran, 1992). Representations can either be snapshots of the real world, or they can express an average state over a certain interval of time. The subject has become increasingly relevant among custodians of ephemerous spatial databases (particularly in meteorology, forestry and navigation) who require consideration of the problems of object identity and changes not only in the spatial and attribute domains, but also in the temporal domain. The addition of the time dimension raises new problems in data structuring (time is topologically unidimensional) and representation. The tools to analyse, generalize and visualize temporal information are still in their infancy.

Model-oriented generalization research has been somewhat neglected in comparison with the efforts invested in graphics-oriented generalization. The traditional view of generalization in support of surveying and mapping organizations for multi-scale map production is overwhelming and has been much more studied. Busy implementing algorithms to perform the analogue of cartographic generalization tasks such as simplification, exaggeration, elimination and displacement, we have forgotten the intimate relationship between generalization at the modelling level and generalization at the 'surface' (e.g. graphical representation). Cartographic generalization requires (1) inside information regarding a spatial object (including spatial, semantic and perhaps temporal aspects), and (2) outside information regarding the relationship among objects and their contextual relevance. The resolution of conflicts, for instance, typifies the problem of generalization on the 'surface', but requires both types of information for its solution. As mentioned earlier, the way the data model is organized and can be generalized is likely to influence the performance of cartographic generalization.

1.5 Cartographic generalization

The tools currently available for automated cartographic generalization resemble those of manual generalization. In this sense, efforts in the automatic domain are oriented towards the manual domain. Furthermore, the quality of computer-produced maps is often tested by comparing the results with manually produced ones. The question is whether we should use manually generalized maps as a criterion of good performance for automated generalization. Should automatically produced maps look like manual ones? This is perhaps a dubious goal and probably unrealistic. Some authors have argued for methods whose results mimic the way people generalize by looking at objects from a distance (Li and Openshaw, 1993). But the fact remains that no new paradigms have emerged under the hat of automated generalization.

A prior attempt towards automated cartographic generalization was to provide a theoretical foundation by answering questions such as what, why, when, and how should we generalize, and providing a framework of objectives to attain, including philosophical, application, and computational ones (McMaster and Shea, 1988). A second step was to make an inventory of the tools available in order to attain those objectives. The list and the definition of those tools vary among generalization specialists, mainly because they fail to differentiate between the transformation applied to an object and the operators used to perform this transformation (Ruas *et al.*, 1993). For example, in the process of simplification, we can list various operators, including select-and-delete, aggregation, compression, smoothing, caricaturization, and collapse. Nevertheless, such inventories were useful since they were used as 'cahier des charges' by commercial firms to set up their development agenda. For instance, a partial catalogue of generalization operators has been already implemented or is intended to be developed by INTERGRAPH, including selection/elimination, simplification, typification, aggregation, collapse, classification, symbolization, exaggeration, displacement, and aesthetic refinement (Lee, 1993). A third attempt was to model the generalization process by suggesting sequential and recursive scheduling scenarios of the generalization steps involving different operators. Those could be different depending on the map subject (Lichtner, 1979; Müller, 1991; Lee, 1992; Müller and Wang, 1992).

One can essentially distinguish between two approaches for the implementation of the working tools in automated generalization. One is batch while the other is interactive.

1.6 Batch generalization

At the most basic level, we have a batch approach where individual algorithms are used to execute various tasks (elimination, simplification, etc.) applied to various kinds of objects. Line generalization has been the most thoroughly studied subject in academic circles (for over 20 years). As with map projections, new algorithms for line generalization keep popping up in the literature. This is no coincidence. Eighty per cent of the cartographic objects are perceived to be lines (in fact, many of them are polygons). Furthermore, single lines viewed in isolation are easier to handle than complex objects like a building or a polygon nesting. Can we now claim that we have reached the state of the art in line generalization? Probably not, especially in view of a lack of theory as to which algorithm is the most appropriate for which line object (river, contour, road, census boundary). Perhaps we need to concentrate more on the application of existing algorithms than on the invention of new ones (Weibel, 1991b). Besides line simplification, we now dispose of algorithms to aggregate and simplify polygons, to exaggerate object size, to collapse complex objects into simpler ones, and to classify and to symbolize cartographic features. But we need an inventory of the performance and the applicability of the different algorithms currently available at universities, national mapping agencies or in private industry. Nobody has a really clear view of what is exploitable. In the generalization tool-kit, however, displacement is not well represented. This is without doubt the most difficult operator to implement, and although some solutions are available (Lichtner, 1979; Nickerson, 1988; Jäger, 1990), they are not comprehensive enough to cover the entire range of possible conflicts. Displacement has become a priority item on the research agenda.

Going one step further, individual batch solutions may be bundled into one 'total' comprehensive batch solution that can be applied for the generalization of an entire map composed of many different objects. Issues such as scheduling management and object interaction have then to be resolved.

A program such as CHANGE, developed at the University of Hannover, is a combination of procedural steps which comes close to the idea of a 'total' solution (Powitz and Schmidt, 1992; Gruenreich, 1993). To develop effectively such a program, one has to define clear objectives. In the case of CHANGE, for instance, the goal was to provide the automated generalization of some feature classes of German topographic maps for a limited range of scales, going from 1:5 000 to 1:25 000. Even in this case, however, the program is suboptimal in the sense that it performs only 50 or 60 per cent of the work. At the end, the user is still required to intervene to perform the necessary quality control and corrections required by operations that could not be entirely automated, such as displacement of conflicting objects in complicated surroundings. As a further example, Nickerson (1988) developed a system for automated generalization of topologically structured cartographic line data. The system is capable of handling feature elimination, feature simplification, and interference detection and resolution. The system is implemented in Fortran, but uses English-like rules for the user to specify generalization options. The intended scale range is 1:24 000 to 1:250 000.

The question is whether a 100 per cent batch solution in generalization will ever be attainable (or desirable). Performance in batch solutions is more likely to follow the economics of 'diminishing returns'. The landscape of geographical features portrayed on topographic maps, for example, can vary almost to infinity. This great variation creates generalization problems which cannot all be foreseen and the research required to cover all cases is so complex and so demanding that it would not be economical. The situation may improve in the future, however, when our methods will be derived from the 'deep' structure (semantic and topology) rather than from the surface level (form and size), and, therefore, will be less sensitive to the variation of individual objects.

1.7 Interactive generalization

The difficulties of providing a batch solution and the disappointment over the progress of the formalization of generalization knowledge (see below) have led some researchers to put their efforts towards the exploitation of interactive techniques. In this case, low-level tasks are performed by the software, but high-levels tasks, such as the choice of an object to be generalized or a particular routine or parameter, are performed or controlled by humans. In other words, the computer implements some tasks (usually execution) which it is good at solving but relies on the user for control and knowledge. Such an approach was suggested by Weibel (1991a) and was termed the 'amplified intelligence approach'. Furthermore, batch technology reflects a line of thought more appropriate to the 60s and 70s than to the 90s. The present trend is to use the interactive environment made available through work stations, PCs and powerful interfaces. So one might say that the dichotomy between batch and interactive generalization is rather artificial and will vanish in the future.

Interactive solutions are based on a user-friendly interface (including multi-window displays, pull-down menus, tool palettes, and menu shortcuts) which allows the user to navigate easily through the system's options and select the objects to be generalized as well as the tools used for generalization. Weibel (1991a, 1991b) gives a detailed list of components required for an 'amplified intelligence' system (mentioned above). Among these are facilities that support the user in making correct generalization decisions (e.g. measures giving data statistics or indicating object complexity; query and highlight functions, etc.) as well as functions for logging of interactions and scripting facilities required. For an interaction approach to be successful, it is essential that it does not just replace the cartographer's pen, but really enables the user to make decisions about generalization on a high level, that is, the system must be capable of amplifying human intelligence. The approach of interactive systems could also be regarded as an equivalent to decision support systems which are frequently used in business and planning applications (Sprague and Carlson, 1982).

The system MGE Map Generalizer produced by INTERGRAPH provides a first step in that direction (Lee, 1993). In a sense, it is comparable to 'electronic' hand generalization, providing more powerful legs to run and a better opportunity to think. The emphasis is on graphic output supervised by human judgment rather than on database or model-oriented generalization. Hence, the approach follows the manual cartographic tradition. Such an approach is more flexible and versatile than a batch program (e.g. allowing reductions along a wider scale range for a greater variety of map types) but the question is whether it is practical in large production environments. The provision of logging and script capabilities which can 'remember' the values of the parameter used and the scenarios that were adopted for similar map situations may give a partial answer.

Interactive systems offer a good potential for the supervised testing and assessment of generalization algorithms and methods. Also, due to mechanisms for interaction logging, they provide an opportunity for the recording audit trails of expert users with the system and thus offer a potential avenue to the formalization of knowledge about generalization processes. Interactive systems have thus also been proposed as a workbench for generalization research. The testing, according to production criteria, of existing batch and interactive methods, or methods which are a mix of these two approaches, must be a priority on our research agenda.

A further degree of sophistication could be reached if we were able to create programs intelligent enough to mimic human thoughts. Cartographic generalization being essentially a creative process (Robinson *et al.*, 1984), it is clear that the batch and interactive solutions need to be combined with some intelligence if we ever want to attain a performance close to a human expert. This explains the growing interest in building rule-based or knowledge-based systems for generalization.

1.8 Generalization and knowledge-based approaches

The introduction of artificial intelligence (AI) in automated generalization is essentially a problem of knowledge acquisition, representation and implementation. The programming languages (Prolog, Lisp) and tools (expert system shells) to manipulate that knowledge and infer generalization decisions already exist. Recursive programming and backtracking techniques, searching strategies and reasoning strategies are part of the problem-solving tool-kit available in any AI software. The fundamental issue is whether we can represent generalization knowledge with 'if–then' production rules that can feed an AI-based system.

Generalization knowledge can be acquired from three different sources: (1) written information available in textbooks and mapping agency guidelines, (2) existing map series, and (3) human cartographic experts.

Three categories of knowledge have been suggested to implement a rule-based system: geometrical knowledge (size, form, distance, etc.), structural knowledge (underlying generating processes which give rise to a cartographic object), and procedural knowledge (operations and sequencing of operations necessary for generalization) (Armstrong, 1991; Müller, 1991).

Partial attempts at gathering information from human experts have been reported (Richardson, 1989). Some knowledge has already been compiled in mapping agency guidelines (e.g. USGS, 1964). A method of 'reverse engineering' is presently being experimented with at NCGIA Buffalo (Leitner, 1993) and the University of Zurich, where existing map series displaying information at different scales are systematically analysed, in order to gather knowledge about generalization.

Observation of map series shows that changes in the graphic representation of an object are sometimes rather abrupt (Ratajski, 1967). In extracting procedural knowledge we have paid too little attention to those 'catastrophic' levels where a change between two successive scales in map series may cause large variations in the representation of the objects (where the polygon envelope of a church turns into a cross symbol, for example). Furthermore, sequencing of operations are usually predefined in already existing batch generalization software (ASTRA, 1986; CHANGE, 1992). But sequencing can also be determined by the user in case the software is interactive (MGE/MG, 1993). In principle, the interactive approach to automated generalization offers more flexibility; it allows the application of procedural knowledge closer to the needs of the user. Its drawback could be that it may require a long period of interactive operations.

The difference between a straightforward verbal account of generalization events and rules is that in the latter case an attempt is made at formalizing the knowledge in some kind of structure which is machine interpretable. The lowest level of formal representation is a look-up table. Imhof's suggested relationship between settlement size, map scale and settlement representation is perhaps the oldest attempt at constructing a look-up table for generalization (Imhof, 1937). But Imhof's table may be also translated into predicate calculus statements, a formal representation of 'if–then' rule statements. Most of the rules available in mapping guidelines can be rewritten in this way. This type of representation leads naturally to programming in logic (e.g. Prolog) at the implementation stage. Another useful representation is the semantic network where the nodes correspond to facts or concepts and the arcs are relations or associations between concepts. One popular application of semantic networks are the hypergraph database structures (HDBS) which describe relations between complex objects and classes (Bouillé, 1984). Similar representations could be used for scheduling and controlling the generalization process.

There is unfortunately little which can be said about the implementation of rules in automated generalization. Except perhaps for name placement (Cook and Jones, 1989; Freeman and Doerschler, 1992), and some experiments to combine procedural and logical programming for the generalization of specific cartographic features (Nickerson, 1988; Müller, 1990; Zhao, 1990; Graeme and João, 1992; Lee and Robinson, 1993; Wang and Müller, 1993) there is no rule-based comprehensive generalization system ready for operation. There is, in fact, no proof that such a system can be constructed. Previous attempts by various NMAs at formalizing cartographic knowledge in the form of a huge collection of rules are rather deceiving. The difficulty comes from the way generalization knowledge presents itself. Most of the guidelines are a collection of common sense statements, addressed to specific cases, such as 'IF windmills appear in large numbers THEN do not show them all' or 'IF contour forms small island THEN do not show'. Those refer to what one might call superficial knowledge (Nijholt and Steels, 1986). But to become operational, a rule-based system cannot rely on superficial knowledge alone. Like a human cartographer, the system should be able to reach deeper knowledge to add to the superficial knowledge when needed. Deeper knowledge refers to more complex reasoning and the ability to make inferences based on geographical context, priority, pattern, map purpose, etc. The real challenge in future research will be the acquisition and formalization of this deeper layer of knowledge which is not in the guidelines but in the mind of the practitioner. Practitioners admit themselves that they find it difficult to rationalize their decisions into a set of formalized rules. Some suggest that deeper knowledge might be extracted through statistical studies of called sequences of procedures and statistical frequency study of calls for each procedure stored in a log-file during the execution of a particular generalization software. It is further hoped that even more specific relationships can be determined using this approach, such as the context in which individual generalization operators may be used. One requirement for research in this direction is the availability of user-friendly interfaces which enable the user to interact comfortably and dynamically with the cartographic software, as if she/he were in a real production situation.

1.9 Generalization and data quality

An almost forgotten consideration in generalization research is data quality. It is obvious that generalization will influence some of the components of data quality, including location accuracy, attribute accuracy, consistency and completeness (Müller, 1991). Displacement will lead to lower local accuracy; completeness will be affected by selection and merging operations; some attributes may be lost through reclassification; consistency may be affected by uneven applications of spatial or temporal abstractions. Note that positional accuracy (i.e. the ability to get access to correct position) is a different issue to shape accuracy (the ability to recognize the 'true' shape of the object). In the former case, the generalized paper map is obviously a poor surrogate to the original database and one might claim that measurements of this sort should be conducted with a GIS, not a map. In the latter case, however, the map remains an indispensable tool. Quality also relates to the specification of the generalized dataset. What was the purpose of the generalization process and how do the results compare to this purpose? Here we need to distinguish between quality in the context of model-oriented generalization and quality in the context of cartographic generalization. Although both objectives are obviously related, they are driven by different purposes. Ultimately, quality is related to fitness with respect to some use: is the data product suitable for the intended use, or conversely what are the possible uses of it?

Generalization may have unpredictable effects on the metrics, topological and semantic accuracies of map products. In a recent study, João *et al.* (1993) showed that the length of a feature usually decreases, but may also increase, with scale reduction. Furthermore, João *et al.* showed that those changes (lateral shift, angular distortion, etc.) may critically affect GIS analysis involving map overlay at different scales. The question here is whether the results introduced by cartographic generalization, a visual-oriented process, could be used in GIS for modelling! On the other hand, model-oriented generalization may have a purpose in GIS analysis. Again the two issues, i.e. cartographic generalization versus generalization for modelling purposes, should not be confused when referring to data quality. Nevertheless, there is an urgent need for a systematic analysis of the effects of model-oriented generalization and the potential dangers of using graphically generalized documents on GIS operations.

1.10 Present and future development: critical thoughts

The following is a critical discussion (at times deliberately provocative and polemic) of what we believe to be the state of the art in automated generalization, with some hypotheses about why we have got so far and yet have accomplished so little, and some postulates regarding possible developments in the future. This section is intended to provide food for thought and provoke further argumentative discussions.

1.10.1 Critical hypotheses

- Most people in cartography and GIS now realize that there is a problem with generalization. Nobody, however, has a clear perspective of what the objectives of generalization should be, and what scale ranges, feature classes, or methods we should concentrate on.
- Nobody in the field has a clear vision of what generalization should be able to accomplish in a digital context. While many researchers argue that generalization should be performed with a different view in the digital domain, most people still resort to cartographic generalization when they claim to be busy developing methods for non-graphic generalization (i.e. model generalization).
- As a result, many researchers confuse the objectives and characteristics of model-oriented versus cartographic generalization.
- In particular, the issue of data quality in the context of GIS and model-oriented generalization is often confused with the objectives of cartographic generalization.
- Most of the research (80–90 per cent) in generalization has focused on rather secondary issues (such as single cartographic line generalization), instead of attacking the burning, more complex issues (such as object-oriented, purpose-dependent generalization processes and database requirements).
- Formalization of knowledge is stagnant because there is too little interaction between the computer experts and expert cartographers. Those who are working on the automation of generalization do not know how to generalize, and those who would know how to generalize are not really being asked (at least not being asked the right questions).
- The data models and data structures which are being used for today's automated solutions are archaic and not capable of supporting any comprehensive approaches involving context-dependent generalization operations (e.g. merging or displacement).
- Limited topological data models (e.g. the commonly used arc–node structure) require on-the-fly geometrical computations for conflict detection. These solutions will never be capable of solving the problem of merging and displacement satisfactorily.
- Achieved research projects in generalization are deliverable in the form of dispersed, incompatible modules developed independently at various places. Hence, a holistic view integrating various generalization processes for a particular scale range and a clear purpose is missing.
- As a result, the few generalization tools which have been implemented in GIS software (like line simplification) have distorted or oversimplified the generalization issue.
- Most research has concentrated on the development of new tools, but little research has been done on evaluation and validation of what already exists. Hence, there is a tendency to reinvent the wheel.
- Academic research has a tendency to turn towards the easiest and most publishable issues that are on the agenda at a given time — rather than turning towards the most relevant and urgent issues (because those might be more complex). One example is the recent interest in user-interface research, which is of little relevance to the generalization problem in its entirety, but in which it is relatively easy to achieve visible results thanks to the prototyping capabilities of hypermedia software (e.g. HyperCard).

1.10.2 Directions for future research

General issues

- Identify the objectives of generalization in the digital context. Identify why and when it is needed (e.g. are there cases where generalization can be or must be avoided, thanks to zooming, multi-window displays, or because of GIS modelling requirements?).
- Develop a specific suite of methods for data abstraction and data reduction (i.e. model generalization) and discriminate clearly between issues of data reduction and quality and issues related to cartographic representation.

Generalization operators

- Identify what are the operations that may be automated in graphical generalization (and which ones cannot), which techniques may be used for that purpose, how and how far can we automate the process, etc.
- Strive for the development of the most complete palette of generalization operators for all types of cartographic features (point, line, area, and surface) and feature classes (transportation, hydrography, etc.).
- Make wiser use of tools that are already in place and assess the applicability (scale range, feature classes, etc.) of existing operators.
- Create test scenarios for existing software and push the operationality of the tools to their limits. Write a set of quantitative and qualitative evaluation specifications in relation to production needs. The question is not what is right, but what is good enough for the purpose at hand (a heuristic question).
- Take a step towards more complex and burning issues. Focus on methods for conflict resolution and feature displacement and for the treatment of point and area features, rather than concentrating on line simplification problems alone.
- Experiment with new approaches, such as neural nets and genetic algorithms.

Generalization and data quality

- Clarify the expectations in terms of data quality both for model-oriented generalization and cartographic generalization. Identify the substantive issues that arise from the application of data-quality criteria as we know them from surveying and GIS in the context of generalized datasets.
- Analyse the potential errors introduced by using digitized generalized maps in a GIS.

Human–computer interaction

- Implement more intuitive ways to interact with generalization operators. For example, provide visual feedback on the consequences of parameter selection; inform the user about possible actions that should be taken; suggest to the user various scenarios for sequential actions and select the one most appropriate for the purpose at hand (Should aggregation follow displacement or vice versa? Should a parameter be involved repetitively, like selection and reselection?). In other words, implement a cooperative behaviour between user and software.
- Provide a pre-generalization report indicating potential conflict areas and designating features which are 'softer' and may be more generalized than others.
- Provide functions that constrain the user in order to avoid actions which do not comply with generalization 'rules'.

- Implement all generalization methods, including batch-oriented ones, in an interactive environment for optimal user control.

Data models and data structures

- Make data richer in terms of the feature attributes they are capable of carrying (concerns data production stage).
- Employ data structures that are capable of explicitly representing spatial proximity and spatial relations for points, lines, and polygons (e.g. Delaunay triangulation, Voronoi diagram).
- Develop methods for the dynamic maintenance of data structures to support changes of feature relations during generalization.
- What kind of data model and data structure are best suited for model-oriented generalization?

Knowledge formalization

- Exploit all available methods for knowledge acquisition (knowledge elicitation, reverse engineering, machine learning, scripting techniques in amplified intelligence systems, etc.) involving both computer and cartographic experts.
- Research cooperation between national mapping agencies (NMAs) and academic research should be intensified. NMAs should state their requirements with respect to generalization functions more clearly, and academic research should take up these issues.
- Likewise, the third player in R & D, software vendors, should be in close contact with developments taking place at NMAs, and sponsor research at academic institutions.

Structural knowledge (structure recognition)

- Methods for the definition and extraction of structural knowledge (structure recognition) are urgently needed.
- This requires a range of supporting functions which are able to express the complexity, distribution, and spatial relationships of cartographic features, and improve the selection and control of generalization operators.

References

Armstrong, M. P., 1991, Knowledge classification and organization, in Buttenfield, B.P. and McMaster, R.B. (Eds) *Map Generalization: Making Rules for Knowledge Representation*, pp. 86–102, London: Longman.

ASTRA, 1986, reported in Leberl, F. and Olson, D., 1986, ASTRA — A system for automated scale transition, *Photogrammetric Engineering and Remote Sensing*, **52**, 251–8.

Becker, B., Six, H.-W. and Widmayer, P., 1991, Spatial priority search: an access technique for scaleless maps, in *Proceedings of SIGMOD 91*, pp. 128–37

Bouillé, F., 1984, Architecture of a geographic structured expert system, in *Proceedings of the First International Symposium on Spatial Data Handling*, Zurich, Vol. 2, pp. 520–43.

Brassel, K. E. and Weibel, R., 1988, A review and conceptual framework of automated map generalization, *International Journal of Geographical Information Systems*, **2**(3), 229–44.

CHANGE, 1992, Internal Report on Generalization Software, Institute of Cartography, Hannover University, 16 pp.

Cook, A. and Jones, C., 1989, Rule-based cartographic name placement with prolog, in *Proceedings of AUTOCARTO 9*, pp. 231–40.

Freeman, H. R. and Doerschler, J. S., 1992, A rule-based system for dense map name placement, *Communication of the ACM*, **35**(2), 68–79.

Goodchild, M.F., 1991, Issues of quality and uncertainty, in Müller, J.C. (Ed.) *Advances in Cartography*, pp. 113–39, Barking, Essex: Elsevier.

Graeme, H. and João, E.M., 1992, Use of artificial intelligence approach to increase user control of automatic line generalization, in *Proceedings, EGIS*, pp. 554–63.

Grünreich, D., 1993, Generalization in GIS environment, in *Proceedings of the 16th International Cartographic Conference*, Cologne, pp. 203–10.

Herzog, A., 1989, Modeling reliability of statistical surfaces by polygon filtering, in Goodchild, M. and Gopal, S. (Eds) *Accuracy of Spatial Databases*, pp. 209–18, London: Taylor & Francis.

Hull, R. and King, R., 1987, Semantic database modelling: survey, applications and research issues, *ACM Comp. Surv.*, **19**(3), 201–60.

Imhof, E., 1937, Das Siedlungsbild in der Karte, *Mitteilungen der Geographisch-Ethnographischen Gesellschaft*, Zurich, Band 37, pp. 17–85.

Jäger, E., 1990, 'Untersuchungen zur Kartographischen Symbolisierung und Verdraengung im Rasterdatenformat', unpublished Doctoral thesis, University of Hannover, 149 pp.

João, E., Herbert, G., Rhind, D., Openshaw, S. and Raper, J., 1993, Towards a generalization machine to minimize generalization effects within a GIS, in Mather, P. (Ed.) *Geographical Information Handling: Research and Application*, pp. 63–78, New York: John Wiley & Sons.

Jones, C.B. and Abraham, I.M., 1986, Design considerations for a scale-independent cartographic database, in *Proceedings of the 2nd International Symposium on Spatial Data Handling*, Seattle, WA, pp. 384–98.

Langran, G., 1992, *Time in Geographic Information Systems*, London: Taylor & Francis.

Lee, D., 1992, 'Cartographic generalization', unpublished Technical Report, Intergraph Corporation, Huntsville, AL, 25 pp.

Lee, D., 1993, From master database to multiple cartographic representations, in *Proceedings of the 16th International Cartographic Conference*, Cologne, pp. 1075–85.

Li, Z. and Openshaw, S., 1993, A natural principle for the objective generalization of digital maps, *Cartography and Geographic Information Systems*, **20**(1), 19–29.

Lichtner, W., 1979, Computer-assisted processes of cartographic generalization in topographic maps, *Geo-Processing*, **1**, 183–99.

McMaster, R.B. and Shea, K.S., 1988, Cartographic generalization in a digital environment: a framework for implementation in a geographic information system, in *Proceedings of GIS/LIS '88*, San Antonio, TX, Vol. 1, pp. 240–9.

MGE/MG, 1993, reported by Lee, D., 1992, in 'Cartographic Generalization', unpublished Technical Report, Intergraph Corporation, Huntsville, AL, 25 pp.

Müller, J.C., 1990, Rule-based generalization: potentials and impediments, in *Proceedings of the 4th International Symposium on Spatial Data Handling*, Zurich, pp. 317–34.

Müller, J.C., 1991, Generalization of spatial databases, in Maguire, D.J., Goodchild, M.F. and Rhind, D.W. (Eds) *Geographic Information Systems*, Vol. 1, pp. 457–75, London: Longman.

Müller, J.C. and Wang, Z., 1992, Area-patch generalization: a competitive approach, *The Cartographic Journal*, **29**, 137–44.

Nickerson, B.G., 1988, Automated cartographic generalization for linear features, *Cartographica*, **25**(3), 15–66.

Nijholt, A. and Steels, L.L. (Eds), 1986, *Ontwikkelingen in Expertsystemen: Hulpmiddelen, Technieken en Methodologie*, Den Haag: Academic Service Press.

Oosterom, P. van, 1989, A reactive data structure for geographic information systems, in *Proceedings of AUTOCARTO 9*, Baltimore, MD, pp. 665–74.

Powitz, B.M. and Schmidt, C., 1992, 'CHANGE', Internal Technical Report, Institute for Cartography, University of Hannover, 16 pp.

Ratajski, L., 1967, Phénomènes des Points de Généralisation, *Internationales Jahrbuch fuer Kartogaphie*, **7**, 143–52.

Richardson, D.E., 1989, Rule based generalization for base map production, in *Proceedings of Challenge for the 1990s: Geographic Information Systems Conference*, Canadian Institute for Surveying and Mapping, Ottawa, Canada, pp. 718–39.

Robinson G. J. and Lee, F., 1994, An automated generalisation system for large scale topographic maps, in Worboys, M.F. (Ed.) *Innovations in GIS: Selected Papers from the First National Conference on GIS Research UK*, pp. 53–64, London: Taylor & Francis.

Robinson, A.H., Sale, R.D., Morrison, J.L. and Muehrcke, P.C., 1984, *Elements of Cartography*, 5th Edn, New York: John Wiley.

Ruas, A., Lagrange, G. and Bender, L., 1993, 'Survey on generalization', IGN/COGIT Internal Report, DT-93-0538, 48 pp.

Salgé, F., Faad, C. and Sclafer, M.N., 1990, Real world description models: the IGN-F BD Topo and BD Carto examples, in *Proceedings of the Workshop 'Towards a Common International AM/FM Transfer Format'*, presented by the AM/FM European Division, Montreux, Switzerland.

Smith, J.M. and Smith, D.C.P., 1977, Database abstractions: aggregation and generalization, *ACM Transactions on Database Systems*, **2**(2), 105–33.

Sowa, J.F., 1984, *Conceptual Structures*, Reading: Addison-Wesley.

Sprague, R.H. and Carlson, E.O., 1982, *Building Effective Decision Support Systems*, Englewood Cliffs, NJ: Prentice Hall.

Titeux, P., 1989, 'Automatisation de Problèmes de Positionnement sous Contraintes: Application en Cartographie', unpublished Doctoral thesis, University of Paris.

Tobler, W., 1966, *Numerical Map Generalization*, Michigan Inter-University Community of Mathematical Geographers, Ann Arbor, Discussion Paper No. 8, pp. 1–24.

Tobler, W. and Moellering, H., 1972, Geographical variances, *Geographical Analysis*, **4**(1), 34–50.

USGS (United States Geological Survey), 1964, *Instructions for Stereocompilation of Map Manuscripts Scribed at 1:24 000*, Reston, VA: Topographic Division, USGS.

Wang, Z. and Müller, J.C., 1993, Complex coastline generalization, *Cartography and Geographic Information Systems*, **20**(3), 96–106.

Weibel, R., 1991a, Amplified intelligence and rule-based systems, in Buttenfield, B.P. and McMaster, R.B. (Eds) *Map Generalization: Making Rules for Knowledge Representation*, pp. 172–86, London: Longman.

Weibel, R., 1991b, Specifications for a platform to support research in map generalization, presentation at Special Session on Cartographic Generalization, 15th Conference of the International Cartographic Association (ICA), Bournemouth.

Woodcock, C.E. and Strahler, A.H., 1987, The factor scale in remote sensing, *Remote Sensing of Environment*, **31**, 311–32.

Zhao, X.C., 1990, 'Méthodologie de Conception d'un Système Expert pour la Généralisation Cartographique', unpublished Doctoral thesis, Ecole Nationale des Ponts et Chaussées, Paris.

SECTION II

Generic Issues

2

GIS-based map compilation and generalization

Scott Morehouse

Environmental Systems Research Institute, 380 New York Street, Redlands, CA 92373, USA

2.1 Introduction

Geographic Information Systems provide a new context for considering problems of map compilation and generalization. There are many similarities between the process of compiling and presenting a digital cartographic database and the traditional map compilation process. Many of the ideas and approaches developed as part of the cartographic design and production process are relevant when using geographic information systems to build databases and visual displays. However, there are some important conceptual differences between the traditional cartographic approach and one based on GIS concepts.

This chapter begins by establishing generalization as a part of map compilation. Then the two models for map compilation and generalization are outlined. The first, characterized as the traditional cartographic approach, considers the physical map as both a repository of geographic facts and as a visual display of the information. The second model, the GIS-based approach, uses a geographic database as the core of the process and separates the compilation of the database from the production of the graphic map displays.

This paper is based on the author's review of map generalization literature: see Brassel and Weibel (1988) or Buttenfield and McMaster (1991) for excellent reviews of this literature. The conceptual framework presented here is similar in some respects to that of Brassel and Weibel. However, here more importance is placed on the role of GIS technology in the context of map generalization.

2.2 The context of map generalization

Map generalization is a general term that is applied to the process of compiling a map from larger scale source maps. The generalization process involves the same decisions regarding map content and graphic presentation that are made in any map compilation process. These decisions, including scale, are governed by the purpose of the map. In this sense, map generalization is no different from other cartographic compilation and design activities; it is simply a name given to a set of techniques we find useful when compiling from larger scale map sources.

In this fundamental sense, there is no real distinction between the 'map compilation process' and the 'map generalization process'. For example, the generalization problem 'Which buildings to choose from the large scale map?' is the same as the compilation problem 'Which buildings to choose from the aerial photograph?' Attempts to develop conceptual frameworks of generalization rules should not be considered as a separate problem from other aspects of map compilation and design.

2.3 The traditional cartographic process

In the traditional cartographic view, the map is the focus of the process (see Figure 2.1). Eduard Imhof said, 'Cartography is the art and technique of map reproduction. It is in a narrower sense to be characterized as a refining process between the original source material and the reproduction' (cited in Swiss Society of Cartography, 1967).

Cartography involves a design phase, in which we consider the purpose of the map and make decisions about content, symbology, and so on. This design phase is followed by a compilation phase, where the cartographer gathers information from source materials and places the information on the map, taking into consideration all of the influencing factors (symbol specification, placement rules, graphics standards, etc.). These two activities, design and compilation, are closely integrated. Design decisions are made in view of the technical and graphic standards for map reproduction and visual communication. The compilation process is controlled by the design.

In this process, the information content, or meaning, of the map is considered simultaneously with the graphic presentation of the information. Graphic design criteria such as 'uniformity of the black–white ratio' are considered at the same time as 'complete representation of major highways'. The decisions which define the content and graphic design of the map are governed by the purpose of the map and by established mapping standards and conventions.

Scale is a central factor in the traditional cartographic process. The map scale, like other design choices, is governed by the map purpose. However, scale controls many other design decisions because it will determine minimum symbol sizes, density of symbols on the map, and so on. These include decisions about both the information content of the map and the graphic presentation of the map (see Figure 2.2).

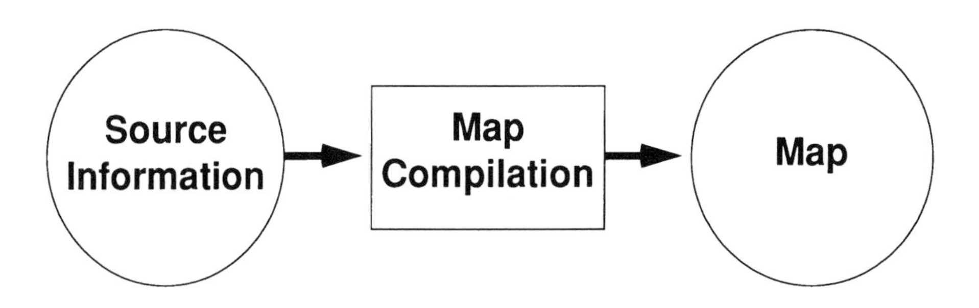

Figure 2.1 Traditional cartographic process.

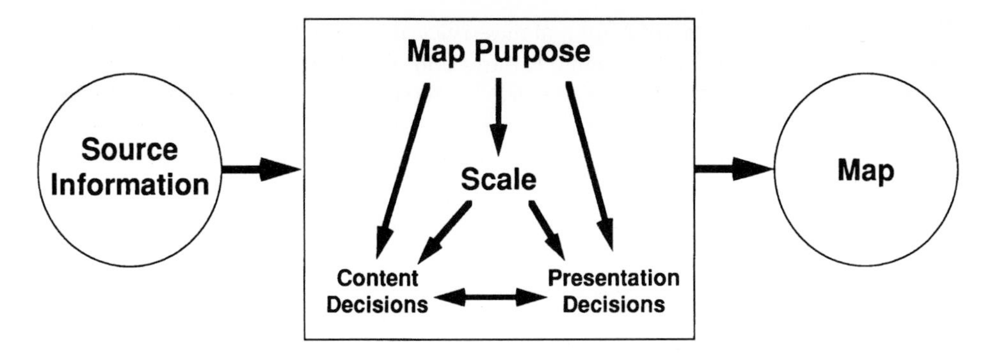

Figure 2.2 *Map compilation process.*

Scale is of central importance as it acts as a filter for the information content of the map. Most graphic presentation issues are based on the characteristics of the printed map surface. The choice of a map scale determines how these factors will influence the map content. For example, a graphic decision on 'minimum separation between linear symbols' will affect the content decision about showing sidings on a railway map. The effect on the map content will be a function of the scale.

Figure 2.3 illustrates the map representation of an urban area at a variety of scales (from Swiss Society of Cartography, 1987). This series of graphics shows that the goal of the map-oriented generalization process is to provide the map reader with the correct impression of the urban structure and to provide a consistent visual look at different scales. Some of the techniques used to achieve this are:

- widths of streets and sizes of houses should remain in the correct proportion to one another;
- maintain a constant black–white ratio;
- delete streets, but indicate their contribution to the urban structure by using rows of houses;
- represent houses along main streets as truly as possible.

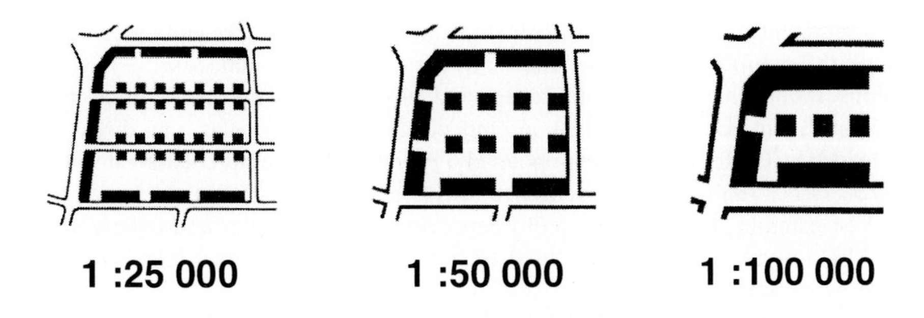

Figure 2.3 *Street and house generalization.*

When considering this series of maps and these rules, it is clear that, at some point, houses stopped being independent geographic features and became signs used to communicate urban structure. The goal of map-oriented cartography is to ensure that the correct overall message is communicated to the mind of the map reader. In this context, the specific map features lose meaning as true objects and become part of an overall visual message. This is possible because the cartographer is simultaneously considering the 'meaning of the map', the features selected from the source materials, and graphic techniques to construct a visual image on paper.

It is difficult to imagine that this sort of cartographic process can be formally defined and automated. The decisions involved in compiling Figure 2.3 are complex and involve an understanding of the map purpose, symbols, reproduction criteria, the map reader, cartographic conventions, and other factors.

The most ambitious approach to automated cartography is to replicate this complex cognitive process directly in software. The cartographic software system would be given a detailed list of rules which define a particular map product — its purpose, scale, symbol specification, graphic tolerances, symbol priorities, rules for depiction of urban areas, etc. The system would then proceed to apply these rules to produce a map, taking into consideration all the influencing factors — selection of the appropriate features from the source material and their representation within the aims of the map product.

Although this approach might be of theoretical interest, this author does not believe that it is of any real practical use for two reasons. First, the complexity of the rules and knowledge involved in cartographic compilation (and generalization) is truly mind-boggling, and is not likely to be realized using current technologies for artifical intelligence. Second, the whole 'map-centric' view of cartography is changing as a result of the geographic information systems approach to working with geographic information.

2.4 The GIS-based cartographic process

GIS-based cartography deals with many of the same issues as traditional cartography. They both involve the collection and communication of geographic information. However, when using a GIS approach, a much stronger separation is made between the information content of the system (the 'geographic features' or 'landscape model') and the presentation of this information in graphic form (the map image or 'cartographic model'). This is shown in Figure 2.4.

In GIS-based cartography, a wide variety of graphic displays (maps) can be generated from a single geographic database. These displays are generated by selecting features from the geographic database, then rendering them at a particular scale using symbols chosen to represent the properties of the feature. As cartographers, we end up compiling two things: the geographic database, and maps derived from this database.

Decisions about which features to collect and how to represent the features in the database usually involve criteria which are independent of a particular map display (or even scale). Many GIS databases are populated by conducting a comprehensive inventory of features, rather than as a representative selection of features for a particular map display. For example, a GIS database of electrical transmission lines and towers would capture ALL towers, not just those 'used to indicate a heavily industrialized area' or 'useful for aerial navigation'.

The distinction between the GIS database, with its associated specification and compilation rules, and the map display, with its specification and mapping rules, is important.

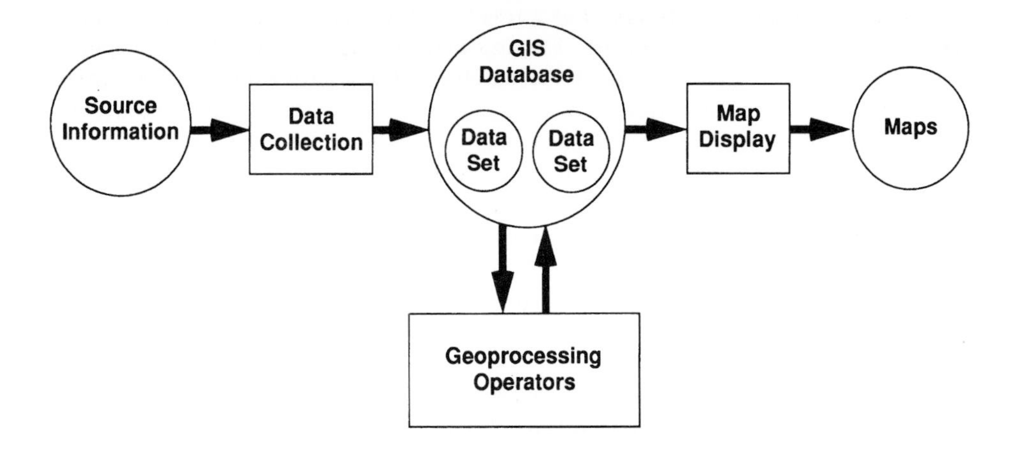

Figure 2.4 *GIS-based cartographic process.*

In this context, cartographic decisions and processes can be characterized as one of three different information transformations:

1. Source to geographic database. These are decisions in defining the geographic data model and associated content standards, as well as the processes useful in populating the database with features. For example, this would include the definition and representation of 'streams' in the database and the methods used to compile them from photography.
2. Geographic dataset to geographic dataset. These are processes which transform a dataset from one form into another together with the parameters which define these processes. For example, coordinate transformation using map projections.
3. Geographic dataset to map. These are decisions in specifying the graphic standards and processes which generate a graphic image (map) from geographic features in the database.

We will consider each of these areas in turn.

2.4.1 Geographic databases

Geographic databases are defined using formal models of geographic information. In defining a geographic database, the emphasis is placed on representing features in a landscape, rather than the graphic display of the features. Most geographic databases are based on a generic model for representing objects in space. Specific geographic feature classes are then realized in terms of the generic model. Examples of this approach include DLG-E, ATKIS, and commercial GIS systems. Each of these systems defines a set of generic objects (features, arcs, nodes, object parts, attribute values, etc.), then utilizes a data dictionary to express the information content of a particular dataset. For example, 'Roads are arcs in this coverage' or 'Roads are objects composed of linear object parts'. Thus, a geographic database is defined by the generic spatial data model together with the specification of a particular database content.

Geographic databases are compiled as a result of a measurement or sampling process which is applied to geographic reality. The measurement process is defined by a specification of the information content of the database. The information content can be characterized by the classification used to measure the phenomenon, the spatial resolution or spatial units by which the phenomenon is measured, and a description of the process used to perform the measurement (see Figure 2.5).

For example, we could represent 'Population' as 'persons per square kilometer', measured by 'province', using the '1990 census' process. This view of data content definition applies to the topographic content of a GIS database as well. For example, we can measure 'water bodies' using a simple categorical classification (lake, pond, etc.) as polygons using aerial photo interpretation with a minimum polygon size of 10 ha.

Geographic objects do not exist 'a priori' in the landscape — they are defined only in terms of a particular measurement framework. This is self evident if we consider statistical maps or maps of biological phenomena. 'Population density' is not an object on the landscape. There are many ways to measure 'Forest type', each of which would result in the delineation of different features. The secondary nature of a geographic feature is less evident for features common in topographic mapping — isn't a 'city' or 'river' a real object in the landscape? When this question is considered more carefully, it becomes clear that cities can appear in a geographic database in a number of quite different ways — as a collection of streets and buildings; as an urbanized area; as a node in a transportation network; as a place name; as a legal boundary; and so on. Rivers have a similar broad set of definitions. If you discuss rivers with a hydrologist, international boundary surveyor, aquatic biologist, and civil engineer you will get very different conceptual frameworks for measuring and characterizing what we casually call a 'river' (see Figure 2.6).

In the case shown in Figure 2.6, we can characterize map generalization as recompilation of the geographic database using a different measurement framework. For example, we can redefine cities as urbanized areas rather than streets and buildings. A GIS provides a set of powerful tools which can be used to perform such shifts in the measurement framework analytically.

• **What Is Measured?**	**Soil Type**
• **What Is the Spatial Resolution or Framework?**	**Soil Mapping Units (Polygons Minimum Size 2 Acres)**
• **How Is It Measured?**	**Soil Conservation Service Standards**

Figure 2.5 Content specification.

Phenomena	**What Can Be Measured?**
"City"	**Street Polygons** **Street Centerlines** **Buildings** **Urbanized Areas** **Place Names** **Node in Intercity Transit Network** **etc...**
"River"	**Shoreline** **Centerline** **Watersheds** **Floodplain** **Soundings** **etc...**

Figure 2.6 Possible GIS database content for different phenomena.

2.4.2 Database to database transformations

Once we have collected data into a geographic data model, it is possible to create new features by applying geometric and attribute transformations to existing features. This ability to create new information using geoprocessing operations is what distinguishes geographic information systems from automated mapping systems. Although much of this functionality has been developed for applications like forestry or land-use planning, this technology can also be used for cartographic applications including map generalization.

Most spatial analysis operations in a GIS work at the level of the generic data model rather than the higher semantic level of a particular content specification. These generic operations can be combined using rules and parameters to 'compile automatically' a new geographic dataset from existing information. For example, a simplified soils dataset could be compiled from a more detailed dataset by applying these geoprocessing operators (see Figure 2.7). In this example, we have combined a thematic transformation (aggregate classes) with topological and geometric transformations (generalize border, for example).

This author believes that many compilation and generalization problems can be best approached using spatial analysis techniques which generate derived sets of features from existing information. This is particularly interesting because we are not only generating a graphic display, we are generating a new, more useful (for some purpose) model of geographic reality. This new set of features can be used for further analysis as well as for visual display. For example, given a set of points representing detailed census enumeration districts and associated population counts, we can analytically derive a population density surface using one of several different techniques. We could threshold this surface at some meaningful value and use the resulting polygons to depict 'populated areas'. We could also use this surface for many other purposes, such as assessing human exposure to radiation (if we had a radiation plume for some event).

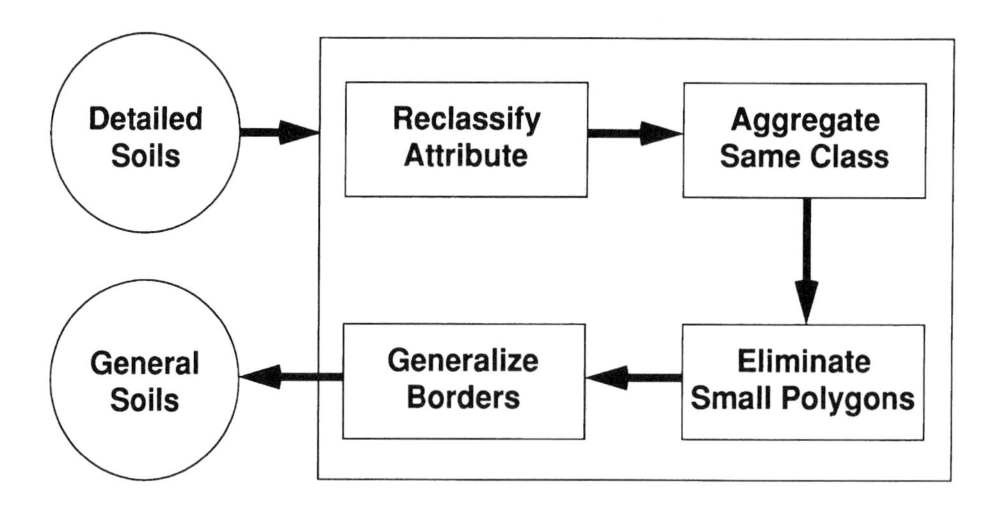

Figure 2.7 Soil generalization process.

We can also use the analytical tools of the GIS to perform transformations which are relevant to the graphic needs of a specific map display. For example, we could generate buffer zones around rivers, then geometrically snap roads which cross inside the buffer zone to follow the border of the zone. This would have the effect of displacing the roads so as not to interfere with river symbols when the features are plotted.

By using the analytical power of the GIS to define new, derived geographic information in the geographic database, we can create simplified features which are more easily mapped. We can also use the GIS to pre-process information for particular map graphics. This reduces the complexity of the mapping process, since a simpler correspondence can be made between features in the database and symbols on the map.

2.4.3 Database to map transformation

In map display, features in a geographic database are translated into symbols on a screen or printed page. The display process is controlled by a symbol specification, which defines the detailed graphic appearance of the symbols as well as the rules for representing features by symbols. These rules can involve feature selection, reclassification, association of symbols with classes, establishing symbol priorities (which symbols appear on top), and the coordinate transformation from geographic space to the map page (see Figure 2.8). These rules can be formalized as a map specification language, which is processed to construct the map graphic. A map is defined by writing a simple program using this language. Mapping rules can also be defined as an intelligent map legend. In this case, a map is defined as a set of themes or information topics. Each theme has an associated data source from the GIS database and a legend defining the classes, symbols, and other rules for depicting the theme graphically. A map can be defined by editing such a legend, then using the legend to render the content of the geographic database.

There are a number of operations which are relevant to generalization that can be applied as the features are mapped using a GIS:

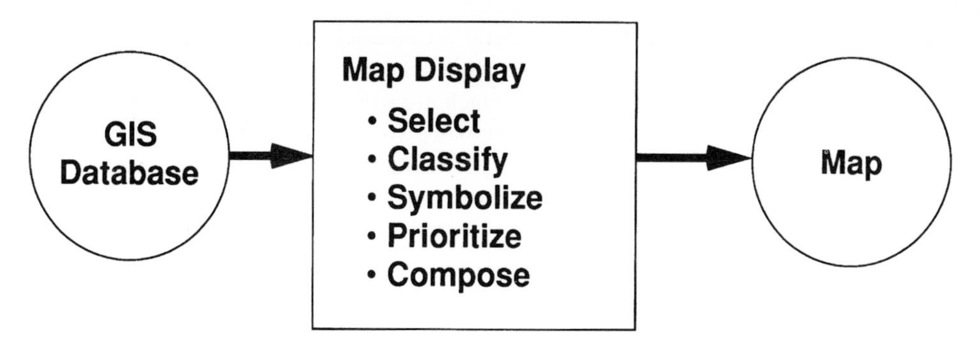

Figure 2.8 Map display transforms features into graphics.

- features can be selected by attribute (for example, all cities which are categorized as capitals);
- features can be categorized by attribute (for example, the number of different road classes could be reduced in number);
- symbols can be generated around feature center lines or placed at a central point in the feature;
- the scale and map projection used to convert feature location on the earth to coordinates on the map display can be set;
- placement of text and symbols on the map display can be adjusted to reduce overlap, based on the geometry of the symbols;
- the coordinate strings which represent features in the database can be simplified or weeded to reduce extraneous detail.

These operations can be applied in different ways to produce many different map displays from a single geographic database. In fact, they can be used to present different information dynamically as a GIS map display is scaled up or down. As the user zooms in on the display, more detailed themes appear and more generalized ones disappear.

2.5 Conclusions

Map generalization needs to be considered in the total context of cartography. Generalization can be characterized as the set of concepts and techniques we apply when compiling geographic information from larger scale source material. As such, generalization issues are a subset of the broader cartographic design and compilation process.

Traditional map-centered cartography is a highly complex and wonderful process in which information is compiled from map sources and represented graphically to communicate some meaning to the map user. This process integrates graphic issues on the visual surface of the map with selection and definition of geographic features from source material. It is difficult to imagine formally defining all of the rules and processes which a cartographer uses in compiling a complex map.

GIS-based cartography separates this process into three parts: compilation of a geographic database according to database schema, processing tools which allow geographic information to be translated from one form to another, and rendering rools which translate objects in the geographic database into symbols on a map.

There are aspects of each of these parts which are relevant to generalization issues. In the initial compilation of the geographic database, we capture information according to some formal model. This model can be very detailed to quite general. We can capture different representations of the same phenomenon as different sets of features. The geoprocessing tools of the GIS can be an extremely powerful mechanism which allows the cartographer to derive new representations of geography from existing information. Such tools can be used to simplify or generalize information, as well as visual presentation on maps. Finally, geographic information can be displayed as symbols on a page or computer display. The process of rendering each database feature as a symbol is quite simple, but the cartographer can define which features are drawn and how they are symbolized using a map specification language. This allows many different map displays to be generated from a single database.

Geographic information systems provide cartographers with many tools and techniques which are useful in accomplishing their traditional goal of map production. They also open new issues related to geographic information gathering and management apart from map production. What is the 'scale' of a geographic dataset? Are multi-purpose, 'scaleless' geographic databases possible or is this just the same as the famous 1:1 scale map? GIS-based cartography provides a useful framework for considering the issues of map compilation and generalization.

References

Brassel, K. and Weibel, R., 1988, A review and conceptual framework of automated map generalization, *Int. J. Geographical Information Systems*, **2**(3), 229–44.

Buttenfield, B. and McMaster, R. (Eds), 1991, *Map Generalization: Making Rules for Knowledge Representation*, London: Longman.

Swiss Society of Cartography, 1987, *Cartographic Generalization: Topographic Maps*, Cartographic Publication Series No. 2, 2nd Edn, Zurich: Swiss Society of Cartography.

3

The need for generalization in a GIS environment

E. Spiess

Institut für Kartographie, ETH Hönggerberg, CH-8093 Zurich, Switzerland

3.1 Introduction

In general terms, this author can agree with the catalogue of problems and questions to be answered as presented in the comprehensive position paper, Chapter 1. On the basis of our group's experiences in map production, both conventional and digital, a point of view is offered here on the need and the benefit of generalization in the production of paper maps, but equally in GIS, and on an approach that seems to be most promising.

From the beginning it should be made clear that we are persuaded that generalization in different degrees and approaches is a must. We are equally persuaded that a 100 per cent batch solution, but also a multi-purpose expert system, is a complete illusion. The mapping or analysis problems to be solved in and with GIS are so manifold, that it seems virtually impossible to take care of all these aspects. Furthermore, when using expert systems there is also a real danger of producing standardized, simplified, rigid and monotonous solutions, lacking the necessary flexibility for each specific case. If we consider the enormous variability of natural and other features in themselves, and in combinations with each other, it is hard to believe that a system could take care of all these incidents.

The purposes and the benefits of generalization are manifold.

1. Applying generalization, we are able to describe reality with different degrees of abstraction, concentrating on the essential information for each group of users.
2. Generalization allows for modelling spatial databases.
3. In order to visualize spatial databases (geodata), some degree of generalization is unavoidable; generalization allows us to enhance visualization.
4. Generalization is the tool to render relevant information legible at a given scale.
5. Generalization allows us to retain the optimal amount of the original information of a spatial database in a given map scale or image format.
6. Generalization allows us to remove noise in an image and enhance the essential parts, not only geometrical but also conceptual noise or redundant information.

3.2 The indispensable need for generalization around GIS

3.2.1 Generalization of digital geodata for the production of paper maps

There is no doubt that we will continue to produce, from digital geodata, paper maps for selected areas, for various paper formats and map scales. A generalization process will be involved, whenever the database has not been purposely structured for the production of such maps, as is the usual case in GIS. The rather firm condition, that most GIS data sooner or later must meet also various mapping requirements, forces the production units to strive for computer-based solutions for generalization. As the primary goal of paper maps is the presentation of geodata in a convenient form for visual analysis, there can be no objection to the request, on the one hand, that the information has to be presented so that the customer can read and decode it under optimal conditions. On the other hand, due to the restricted accuracy of such paper maps, performing quantitative analysis on paper maps can be of only limited importance.

The production of always up-to-date multiple scale versions of maps on the bases of scale-independent databases, with one single maintenance procedure, would require a huge, well-conceived and consistent system of efficient generalization facilities. The updates have to be inserted and adjusted to the geodatabase. This is by no means a straightforward undertaking for an automated system, as may be illustrated by Figures 3.1–3.3. The example has been extracted from the material of the Fribourg test of Commission D of the OEEPE on updating topographic maps by photogrammetric methods (Spiess, 1985). Figure 3.2 is a graphic record of all the changes that have to be made on the original data. They involve crucial decisions in a rather complex adjustment procedure. It is hard to believe that the operator can take care of all considerations which are necessary for deriving generalized updated smaller scale maps and consistent with everything his predecessors have implemented in the original dataset.

We should not forget also that generalization has already begun in the interpretation phase. Figure 3.4 illustrates what eight different operators restituted from the same aerial photograph in the same OEEPE test. To some extent this result stresses the need for field identification. But where houses are concerned, there is mainly a difference in the generalization philosophy of the individual centres. Different minimal dimensions are used and intricacy of detail varies considerably. There is no general denominator for simplifying the forms of the same houses. It seems to be quite a problem to formulate rules for this very common kind of undertaking.

For the production of small-scale thematic maps at various scales for the Swiss National Atlas another approach has been followed: we found it necessary to create a set of digital base maps that correspond to the conventional ones we have made use of for three decades already. Among others, we have created a whole set of digital files on the basis of the existing films 1:1 100 000 which can be used to design base maps at the scales 1:800 000, 1:1 000 000, 1:1 100 000 and 1:1 250 000, but each of them with different symbolization. The originals have been scanned; the data was vectorized and cleaned. The package contains nothing more than the features hydrography, boundaries and hill shading, which may be used in combination (Figures 3.5 and 3.6). But it consists of more than 30 files in order to meet all forseeable combinations, so, for example, the river network is in three different densities, each one with up to nine classes of line widths, which allows us to taper the lines from the source to the mouth. Each of the four hierarchical boundary systems is available in two versions, entirely closed boundaries or interrupted where the river substitutes the boundary. Islands and exclaves, but also the map frame and the

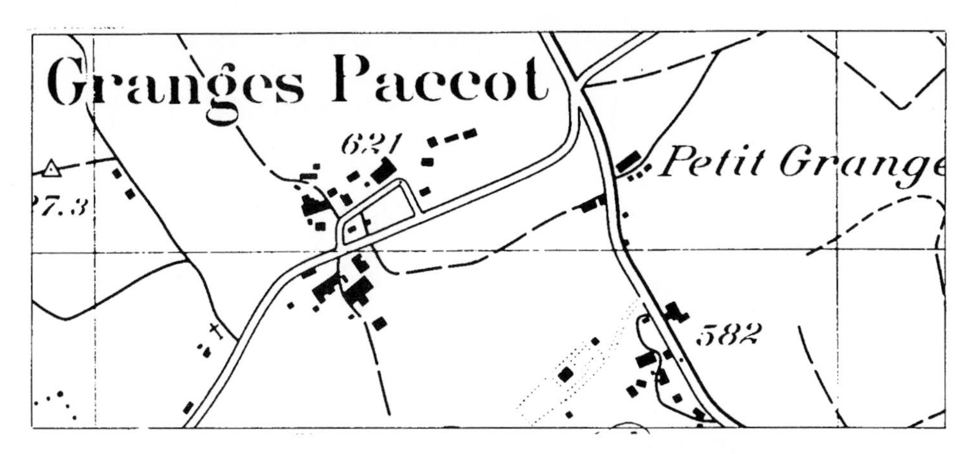

Figure 3.1 *Topographic map at 1:25 000 scale as of 1968, enlarged to the scale of 1 : 10 000.*

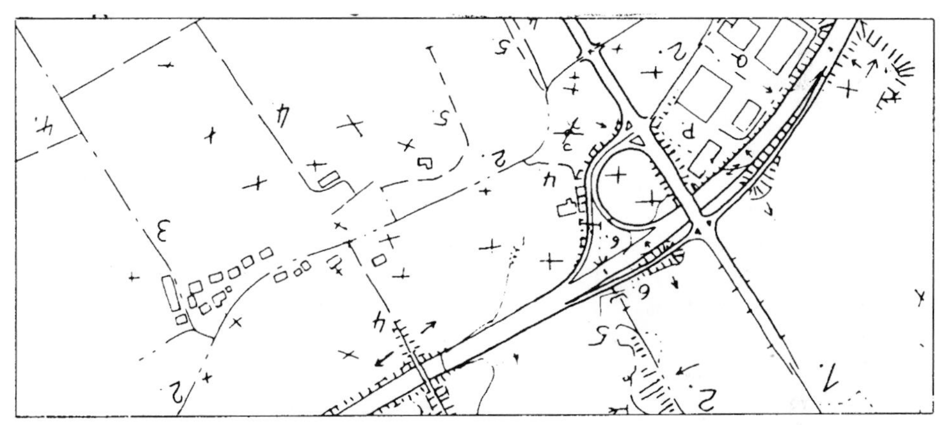

Figure 3.2 *Manuscript of updates; deletions (×), changes and additions, enlarged to the scale of 1 : 10 000.*

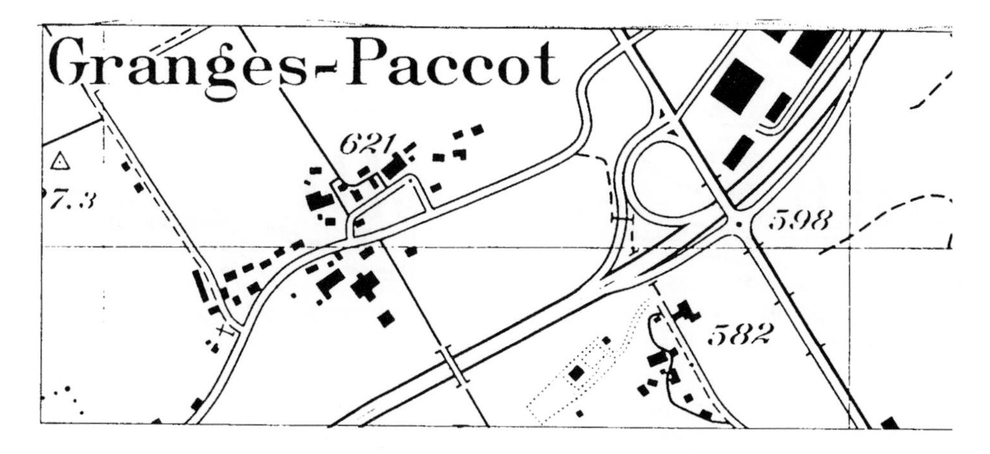

Figure 3.3 *Updated topographic map as of 1975, enlarged to the scale of 1 : 10 000.*

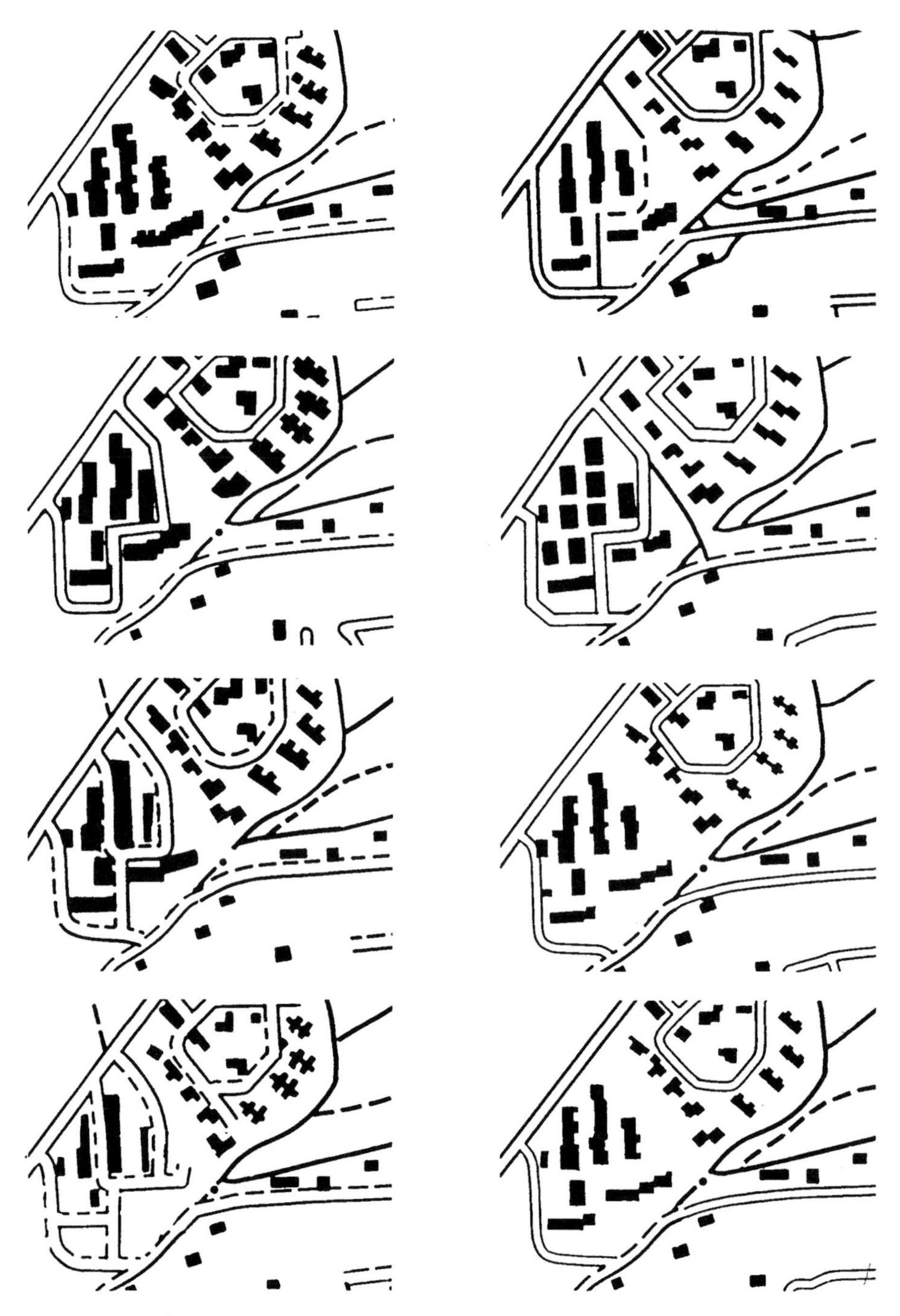

Figure 3.4 Different generalization of houses in restituting the same aerial photograph in the Fribourg test of Commission D of the OEEPE.

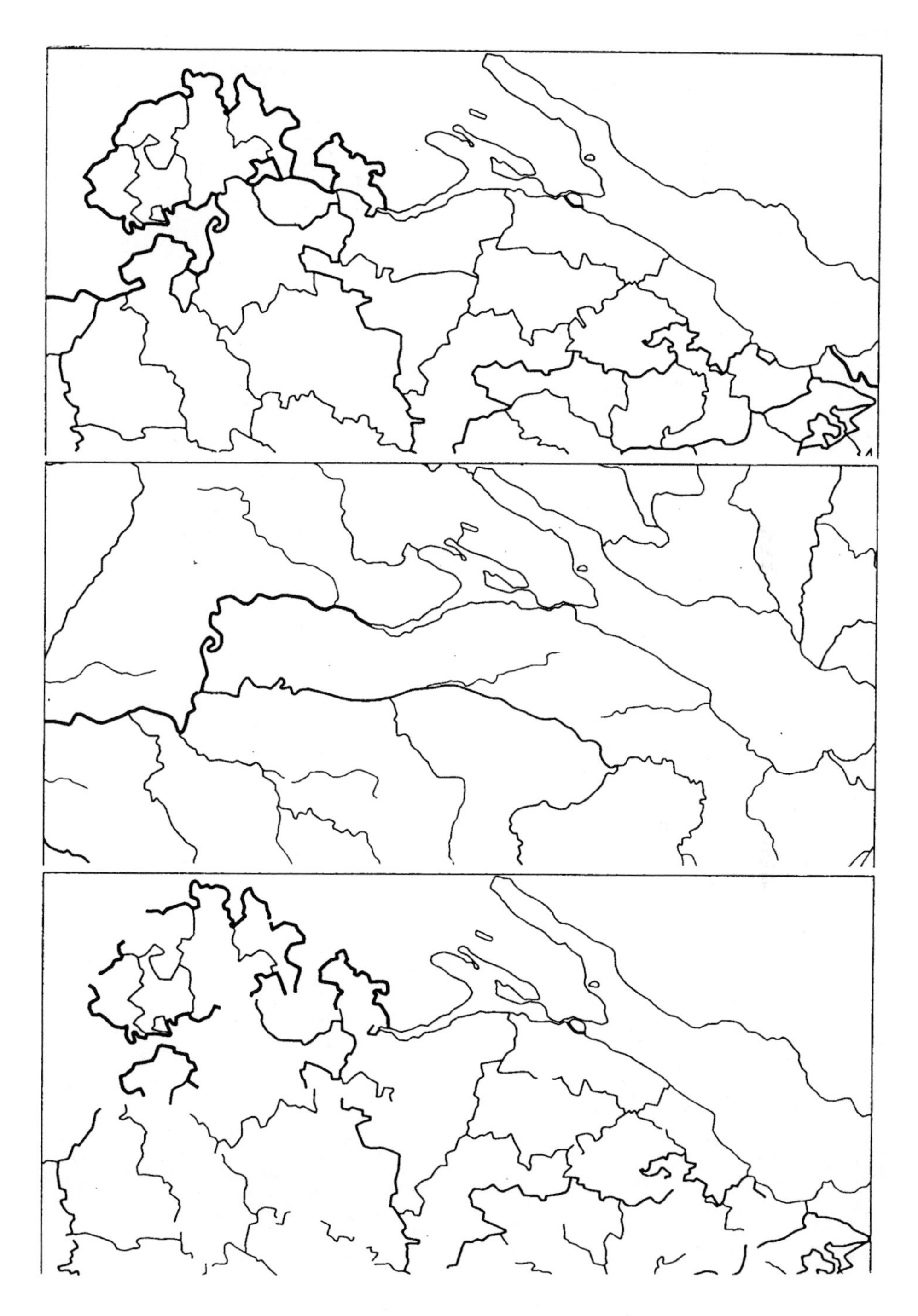

Figure 3.5 A set of coordinated digital base maps for thematic maps; hydrography and interrupted district boundaries.

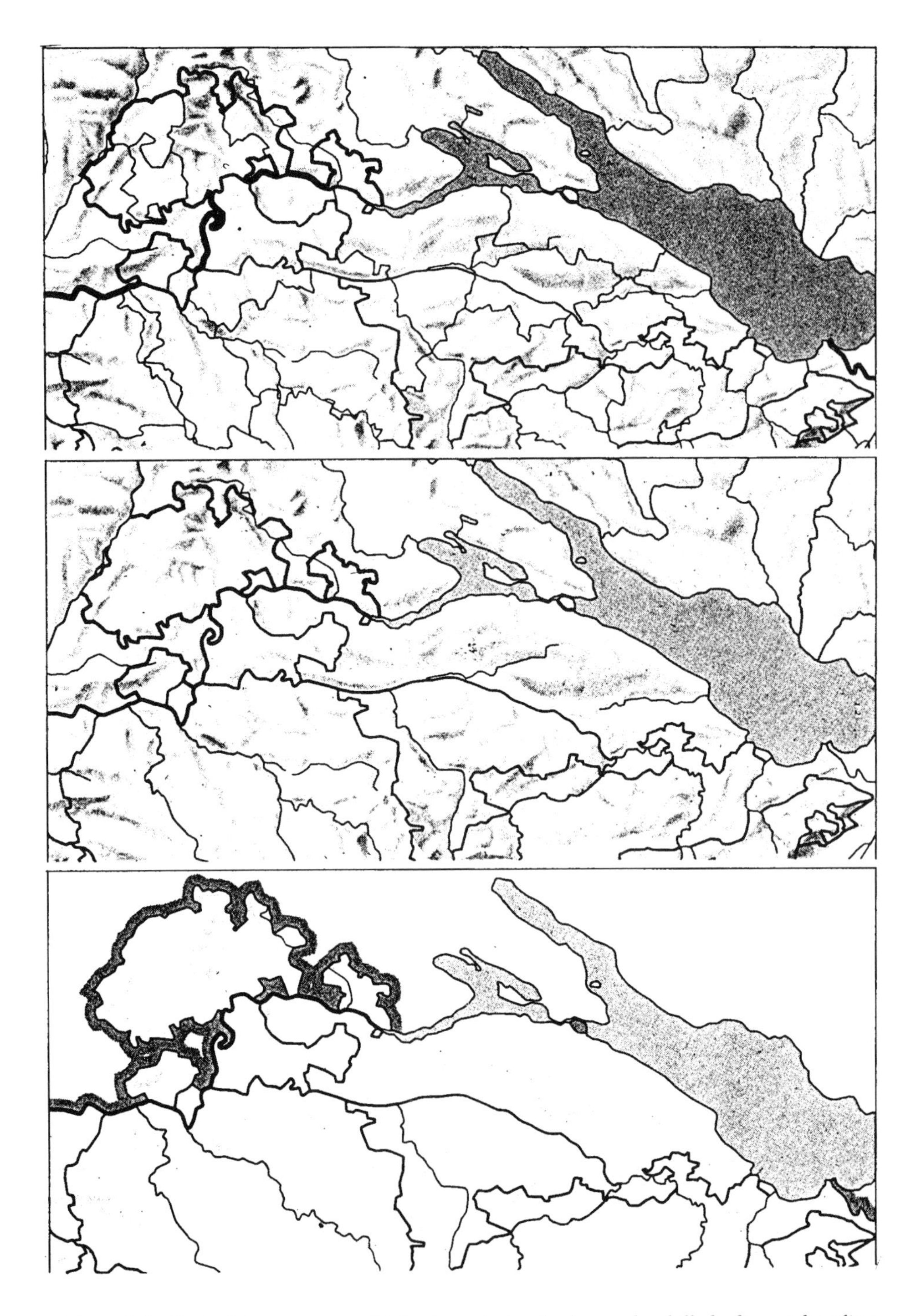

Figure 3.6 Digital base maps combining boundaries, hydrography, hill shading and outline masks.

register marks, are treated with special files. Two versions of hill shading are ready for combinations; others may be derived according to specific needs from the scanned COT images.

Great importance is attached to the requirement that all visual and geometrical conflicts between the different feature types are solved when they are combined. We wonder how a system could be taught to make, for example, all the necessary adjustments between rivers and boundaries as marked in Figure 3.7. Once this work is accomplished, we are left with the updates that are inserted interactively. Fortunately, they are rather seldom in these small scales, but nevertheless the effort must not be underestimated, as several of the 30 files may be involved.

Visualization of GIS data and results of GIS analysis in the form of paper maps will continue to be a demand, as not everybody has access to a workstation or the necessary background to use it. Certain groups of clients may wish to get such map products on their table, and to include them in written reports. This means that they can no longer rely on the zooming facility of the workstation but have to study an analogue graphic image of sufficient quality. There will be a request for even higher quality if the maps are to be printed. In one way or another generalization cannot be avoided in these cases.

In a number of recent projects we have been confronted with indigenous data from GIS which had to be overprinted for final publication on slightly generalized topo sheets (Brandenberger and Spiess, 1993). Overprinting them as they were would have created more confusion than clarification. Considerable efforts, including generalization, were necessary to adjust the GIS features to the map image, so that the message came through (Figure 3.8). In this respect not only topographic map data must be considered, but numerous other topics. This issue will be discussed in more detail later on, when the

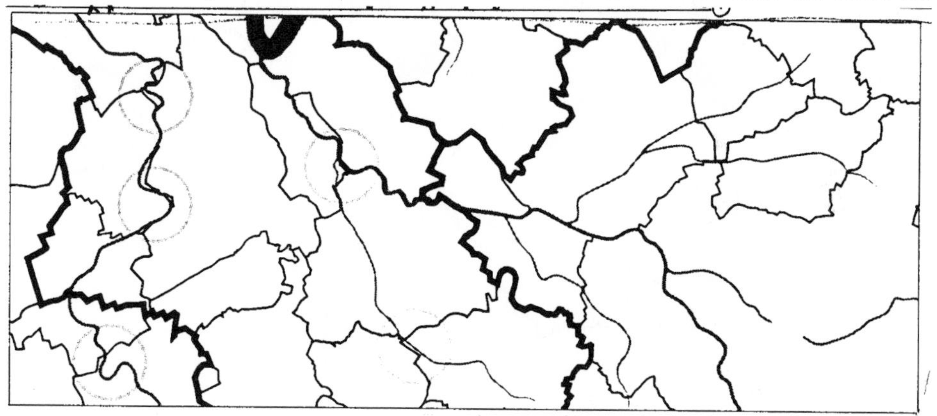

Figure 3.7 Conflicts between hydrography and boundaries.

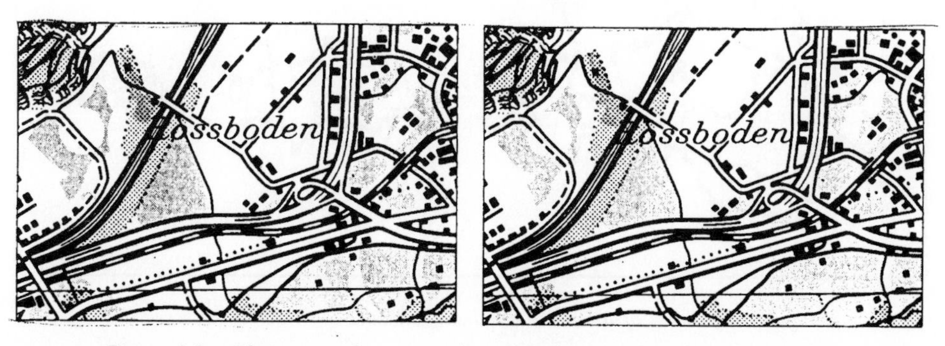

Figure 3.8 Obvious need to generalize GIS data overlayed on a base map.

question of whether generalization is also needed for presentation of data on the monitors of GIS has been answered.

3.2.2 Generalization of digital geodata for presentation on screen

Zooming out on screen for location

When a whole large dataset is fitted to the window or a limited part of the window of the display, usually part of the data is lost. The practically infinite number of discrete x,y-locations of the stored geodata shrinks down to some 1024 × 1258 possible pixels on the screen. The effect is that of a scale reduction. What do we intend then by selecting this option? Often we may simply wish to get a general overview in order to locate an area to zoom in on in the next step. But is such an image, e.g. the last step in Figure 3.9, really the optimal solution to find the source of the Rhine in a dataset of Europe? It is just a quick solution, but is far from being satisfactory for an operator, even if he knows his geography well. Nobody hesitates to complain about overhead or slide projections with too tiny details, but apparently people seem to accept more easily illegible views of the full

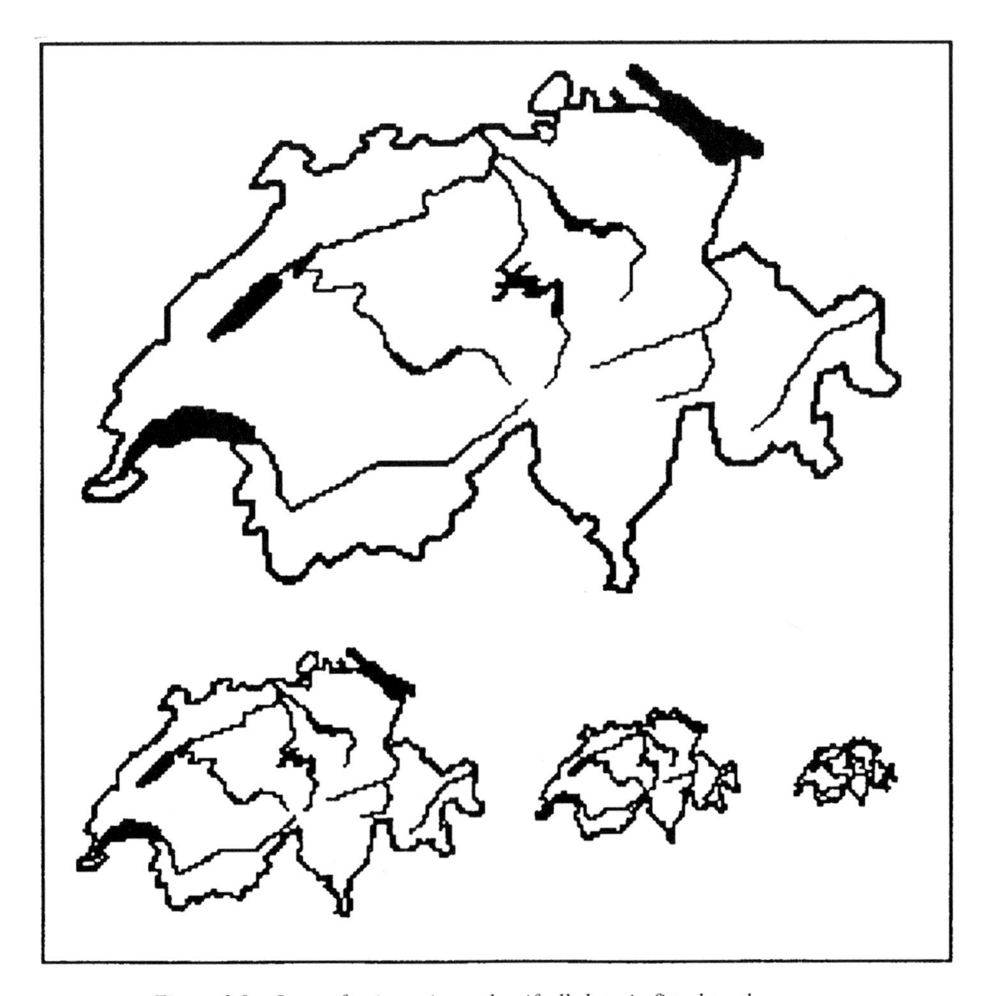

Figure 3.9 Loss of orientation value if all data is fitted to the screen.

dataset on the display, which are far from ideal for interpretation. It is probably accepted only because one can quit quickly this awful image by zooming in again. It is only logical that this deficiency of the total image led to the requirement for fast-scrolling facilities.

Minimal dimensions of data displayed on a screen

Minimal dimensions are rather coarse on a screen in comparison to a paper map with line widths of 0.1 mm. A pixel on a 19 in. monitor (385 × 287 mm) with 1184 × 884 pixel resolution measures 0.28 mm. This causes rugged lines (see Figure 3.10), an effect that to some extent may be mitigated by the aliasing technique. Furthermore, the colour con-

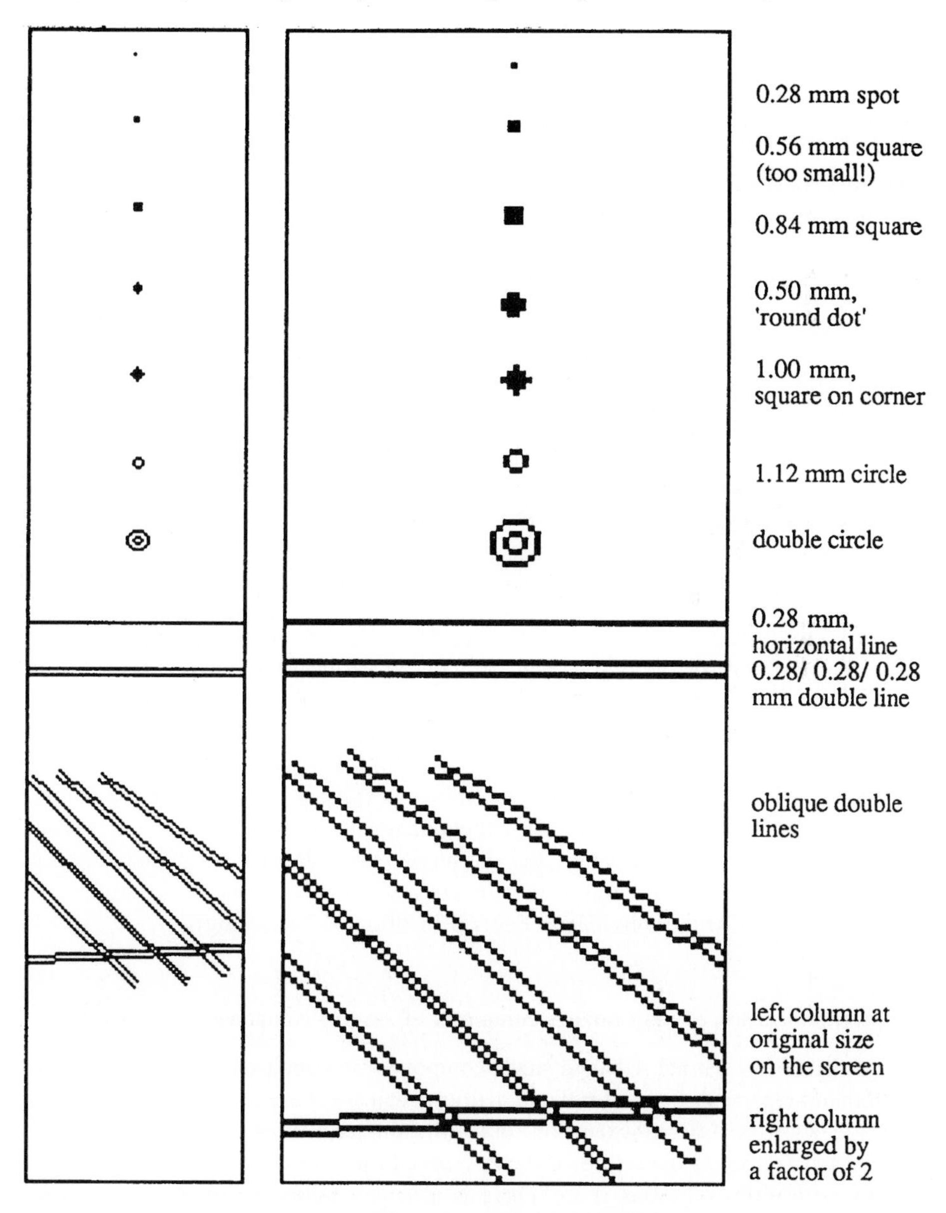

Figure 3.10 Minimal dimensions on a CRT screen.

trast, that can be established for images on a display, is often not as strong as on paper. In a number of tests it has been shown that for clear identification of a topographic pixel map on such a screen, the paper map has to be enlarged at 250 per cent. In other words, on such a display we can work only with sections of 12 × 10 cm, spread over the whole screen, of any scanned paper map. This means that we have to split up a topo sheet for inspection into 30 segments. Any map image larger than 12 × 10 cm and displayed as a whole on the screen at its original, or even reduced, scale, lacks identifiable information and produces instead nothing but noise. The only remedy for such a situation is generalization.

The overall image — a must for certain analysing tasks

We should not forget in this context about the need for an overall view of the entire map for quite a number of other map interpretation tasks. For all general questions the eye has to scan over the whole area, as, for example, recognizing:

- the most congested areas,
- the highest concentrations,
- all empty spaces,
- the total evidence of a certain feature,

or comparing

- regional densities,
- distribution patterns, etc.

It is this kind of interpretation task, which can be performed most efficiently by humans. But they cannot be successful if the overall image is illegible. Scrolling around is often a means to overcome such deficiencies, but is not appropriate for this purpose.

Comparison of vector and raster data that are scaled down on a screen

After all, vector data are easier to present when the scales become smaller (Figure 3.11), because the attributes of the objects can be changed easily. But some effort has to be made to remove those features that are not absolutely needed and to improve on the symbolization. Nevertheless, the image collapses when the scales become considerably smaller.

The same image in raster format is even more sensitive to scale reductions (Figure 3.12). Zooming out fills up the open spaces or eliminates small elements. There are certain solutions to improve on this situation, but the success depends on a thorough analysis of each individual image. An attempt in this direction was made by one of the students at the Department of Cartography, Swiss Federal Institute of Technology.

3.2.3 Generalization of map images composed of several components

No generalization is needed if only a single component is visualized (Figure 3.13(a)). In this woodland representation natural and artificial structures are visible and distinguishable, as are densities, different extensions of forests, different types of forest, etc. But this cannot be satisfactory as the results of the interpretation can neither be localized nor are there any indications why this is so. There is always a minimum of referencing data needed, e.g.

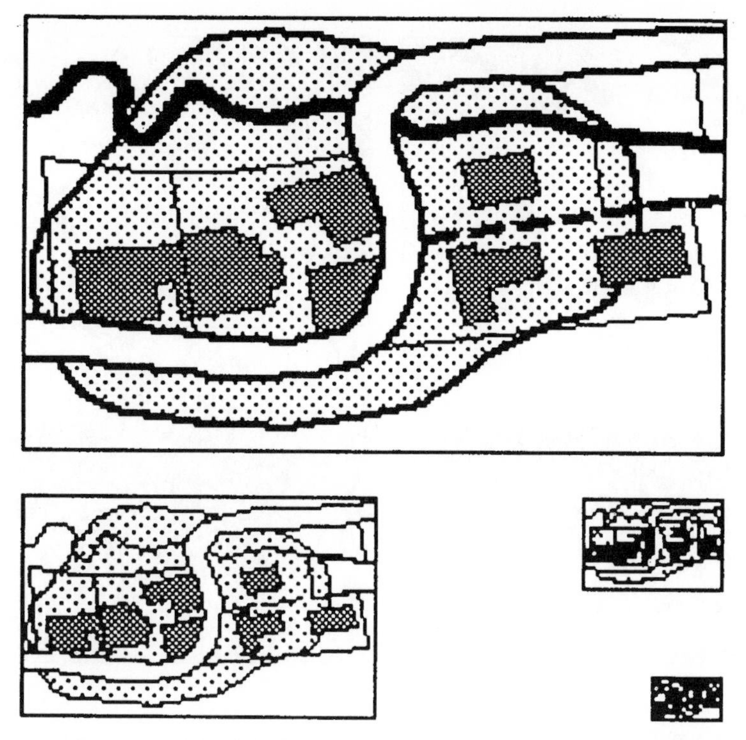

Figure 3.11 Simulated screen images; collapsing vector data due to scale reduction.

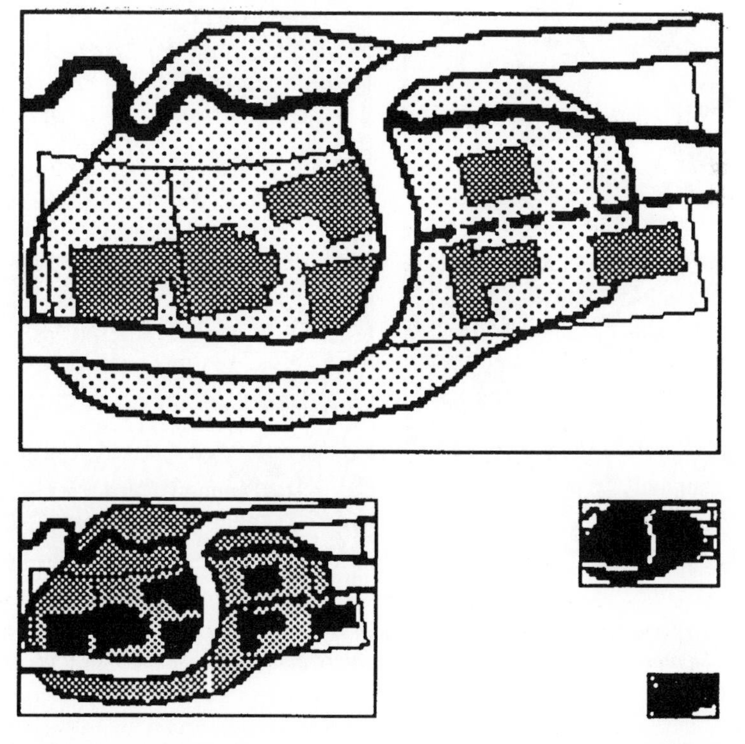

Figure 3.12 Simulated screen images; collapsing raster data due to scale reduction.

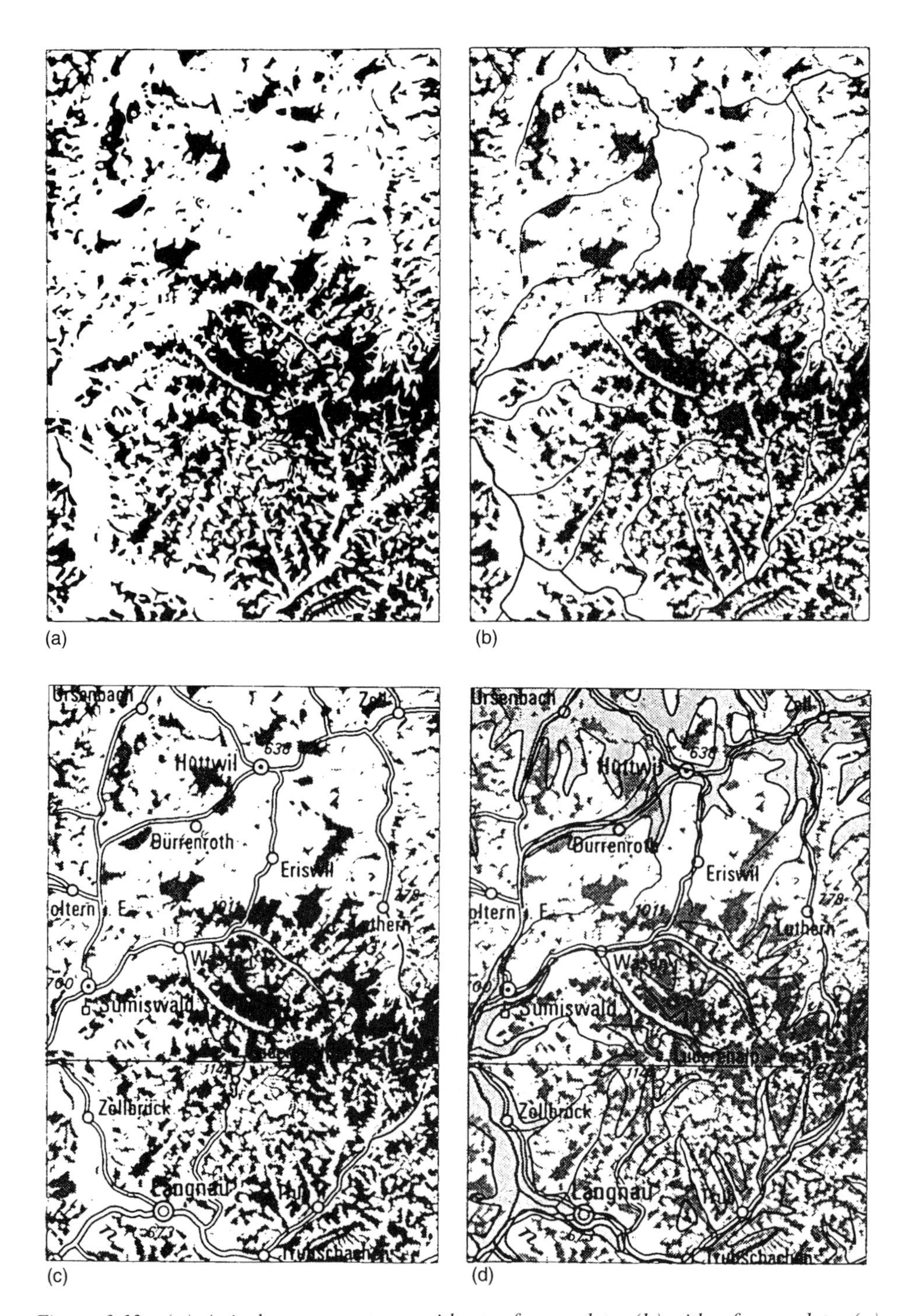

Figures 3.13 (a) A single component map without reference data, (b) with reference data. (c) and (d) The more additional components the more generalization becomes a necessity.

- the main river network; however, not an exhaustive hydrography plate (Figure 3.13(b)),
- names, but only those pertinent to the topic (a set of names can be provided for referencing, but not before the topic is known (Figure 3.13(c))),
- in some cases hill shading, but not necessarily at the scale of the forest,
- or additional components that correlate with the phenomena (Figure 3.13(d)).

In many cases one will not yet be satisfied when the data are referenced, but ask about possible correlations with other components, e.g. elevation, road access, tectonics. People will wish to know why the woodland areas have such different structures.

Therefore, there must be room in the map for additional components or features. But no doubt this is impossible without solving overlap conflicts. Furthermore, as the other components become more and more important for the analysis, the original woodland image will be much too noisy and strive for generalization. Why should we hesitate then? A good generalization enhances the characteristics, which might even facilitate the analysis over the whole image. On the other hand, we can zoom in anytime to the basic data for studying the structures in detail.

3.2.4 Model-oriented generalization

Feature selection and symbolization for the screen presentation of models

One might go as far as to pretend that every display of geodata is preceded by a generalization if some kind of selection out of a total GIS dataset has occurred, selection and elimination being generalization procedures. Even deciding on the symbolization of the features of a model to display may have a generalization aspect as well, e.g. in exaggerating or emphasizing a line by its line width in relation to other features. Both of these measures might be easily controlled by procedures. The criterion being the total 'load' of the image area that can be accepted for legibility, it should be no major problem to determine it. The other demand will be directed by priorities attached to the data and selected when such a display is initiated.

Data aggregation, another model-oriented generalization process

Aggregation of individual data within larger reference units is another case of model-oriented generalization. As an example, we may refer to statistical maps on employment in various industrial branches. In the Atlas of Switzerland at the scale 1:500 000 every community having more than 100 employed persons can be represented. This, however, causes congested areas around larger cities (Figure 3.14(a)). For the school atlas at 1:1 600 000 scale, the smallest unit is equal to 250 employees, thus necessitating regional aggregations of smaller neighbouring communities with similar structures (Figure 3.14(b)). In addition, the number of branches within the pie diagrams has been reduced.

These constructions have been realized with user-programmed modules alternating with interactive phases, e.g. for the selection of communities to be aggregated and for moving diagrams around to a location which is optimal for legibility (Hutzler and Spiess, 1993). The choice of the type of diagram (from a large number of alternatives), the final figure scale, the parameters and the class intervals for all numbers that were too small for a graded circle, are examples of a series of interventions by the operator within the whole production process. One may argue whether these procedures should be considered as model-oriented or cartographic generalizations, as some of these decisions are condi-

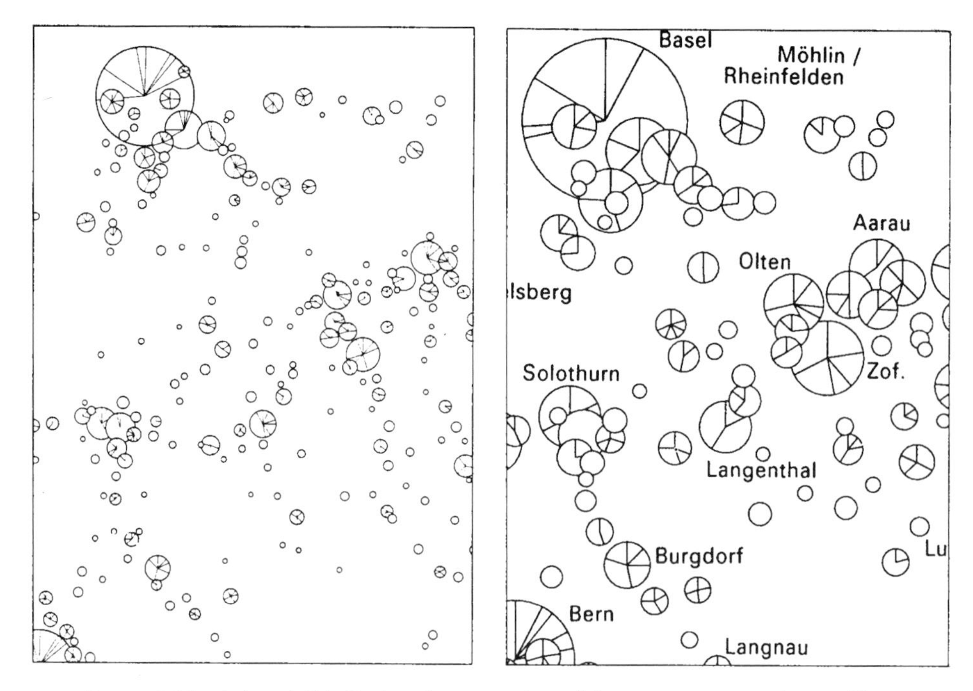

Figure 3.14 (a) and (b) Regional aggregation of data, a means to solve conflicts.

tioned by purely graphic constraints. On the other side they have a considerable influence on the model by their higher degree of aggregation and abstraction in concentrating on major branches only. A case that causes serious problems in generalization (Spiess, 1990c) concerns aggregating a land use pattern, if its components fall under the threshold of surface size.

Of course, one may access within the GIS, at any time, the original data with full precision and accuracy. The above generalization measures have to be taken for overall view analysis and in case of a need to visualize the respective total image.

The importance of external information in most generalization processes

The example given may serve also as an illustration of the observation that decisions in generalization often depend on external information or facts. The delimitation of a region here not only depends on neighbourhood and similarity in production structures, but also on other attributes of a common economic region which may not be part of a GIS or may need a specific assessment as to their correlation to the facts represented or to possible issues of the map interpretation. Another typical example is the construction of isotherms based on the data of a couple of weather stations in a mountainous area. There are contouring packages available for this purpose, for interpolating isolines between the given values. But additional knowledge about temperature/elevation gradients is indispensable for a sound construction (Spiess, 1990a). In a map on population development in the period 1980–90 we have classified 3024 communities with the help of a cluster algorithm. There were some 20 communities with major irregularities, difficult to classify. Again, external information, in this case on the history of railway and power dam construction, helped to find an acceptable solution (Spiess, 1990c, pp. 26–7). It is along these lines, among others, that we justify our reservations as to a generalization machine.

3.3 Interactive approach with knowledge-based modules

We have many reserves in expert systems working with a top-down strategy. Is the system user ever able to specify his requirements so that the product will meet them? As such a user interface could certainly not be a simple one, but instead encompass many questions, the user would have no idea about the impact of his answers. The system, therefore, will not fully understand the problem, a prerequisite to solving it accordingly.

Furthermore, the extreme variability and complexity of the initial data leads to an endless series of rules, constraints, and algorithms with numerous free parameters to compete with in order to handle successfully all possible cases. However, we can imagine that a number of intermediate steps can be formalized, sooner or later, when the huge task is broken down in subunits that are clearly defined. The result of these generalization modules must be evaluated visually step by step and not only the final image. This allows us to gain experience and expertise in the effects produced by certain parameters or algorithms on a specific occurrence of data.

Following the scheme shown in Figure 3.15, we present the idea that the structured initial database should be complemented by a number of generalized versions, especially

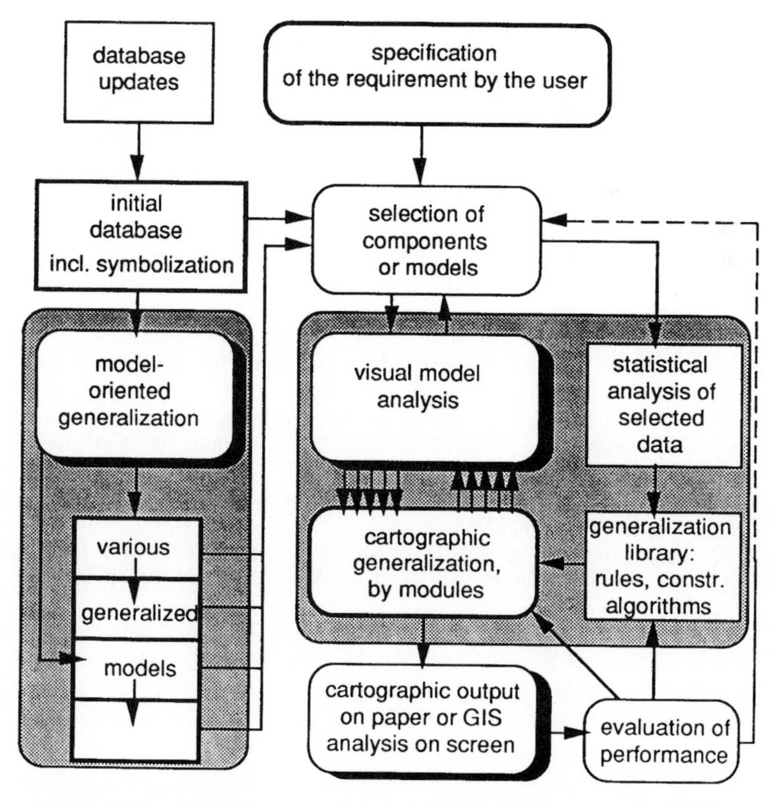

Figure 3.15 Generalization scheme making use of a series of modules.

of data components frequently used for referencing other thematic data. It would seem to be a waste of time to undergo each time all the complex procedures necessary to create a generalized set of base map elements, the more so as we may have to cope with some amount of interactive refinement. Of course, such a set of derived models has to be updated periodically or on demand. This approach seems to be worthwhile, especially when hardcopy output is envisaged, but also for GIS analysis under some degree of higher abstraction.

The system user selects the object and reference data, including an initial minimal symbolization, and analyses the combination visually on the display in view of the following generalization process. For quick scale changes he should have at his disposal an image controller in the random access memory. In parallel, and as a complement to this visual inspection, we propose a series of modules for a statistical analysis of the selected data, again with respect to the generalization strategy to follow. By this scanning procedure we obtain statistical data about the specific properties of the dataset chosen, which can be used as input for the cartographic generalization.

The generalization procedure consequently is broken down into a sequence of modules. These modules are contained in a library and may consist of rules, constraints and algorithms. Each step is verified on the display and repeated with improved parameters if necessary. The critical element is, of course, the sequencing of these modules and their interdependencies. If the performance of a whole scheme, a single module or algorithm, or a set of parameters is good under the given conditions, they may be used to improve on the content of the library (for an application see Hutzler and Spiess, 1993).

Such an approach seems useful, especially in the development environment, but may prove to offer more insight in generalization effects, more creativity and therefore more acceptance by the system users, and also give more flexibility in production.

References

Brandenberger, Ch. and Spiess, E., 1993, From GIS to final print-ready films, in *Proceedings of the 16th International Conference on Cartography*, Cologne, 1992, pp. 721–34.

Hutzler, E. and Spiess, E., 1993, A knowledge-based thematic mapping system — the other way around, in *Proceedings of the 16th International Conference on Cartography*, Cologne, 1992, pp. 329–40.

Spiess, E., 1985, Revision of 1:25 000 topographic maps by photogrammetric methods. Final report of the results of the FRIBOURG test, carried out by Commission D of the OEEPE. *OEEPE Official Publication No. 12*, Frankfurt a.M., OEEPE.

Spiess, E., 1990a, Generalisierung in thematischen Karten, in *Kartographisches Generalisieren, Kartographische Schriftenreihe Nr.10, Schweizerische Gesellschaft für Kartographie*, Baden, 1990, pp. 63–70.

Spiess, E., 1990b, Bemerkungen zu wissensbasierten Systemen für die Kartographie, in *Vermessung, Photogrammetrie, Kulturtechnik* 1991/2, pp. 75–81.

Spiess, E., 1990c, Kartengraphik und Kreativität auch mit digitalen Daten und Technologien, in *Wiener Schriften zur Geographie und Kartographie, Band 4*, Wien, 1990, pp. 23–38.

4

Development of computer-assisted generalization on the basis of cartographic model theory

Dietmar Grünreich

Institute of Cartography, University of Hannover, Appelstrasse 9a, 30167 Hannover 1, Germany

4.1 Introduction

Models are tools of various disciplines. They are a prerequisite of communication in and between the disciplines. A model is made up by purpose-oriented classification and reduction of the original. Every modelling process is primarily a *generalization process.*

Since the 60s, cartographic model theory has developed on the basis of information theory, communication theory and semiology. Cartographic model theory has its roots in the fundamental human concept of mapping, i.e. the activity of every human being to secure their existence by defining their location in space and the accessibility of the spatial environment. Every human being develops individual concepts of space in mind, called cognitive maps. If cognitive maps are to become a common base of social action, then they must be translated into forms which allow interpersonal communication. This so-called *societal mapping* is the subject of cartography. During the development of man, graphic maps became tools of spatial cognition, spatial organization and spatial action. They proved to be the best and most versatile representation of mental maps (Freitag, 1987).

Cartographic model theory describes the communication process related to geoinformation using three types of models of the environment:

1. **primary models** provided by the disciplines involved,
2. **secondary cartographic models** designed for visual communication and spatial reasoning, and
3. **tertiary models** resulting from the thought processes of the map users (cognitive maps).

Figure 4.1 presents a comprehensive overview of the relations between those models. It is called the 'Cartographic communication network'. Furthermore, in context with organizational, hardware and software aspects the diagram is dealing with the set-up of a cartographic information system which can serve as a model for geoinformation systems.

The fundamental changes of communication technologies from traditional cartographic techniques to GIS technologies has changed societal mapping. However, the

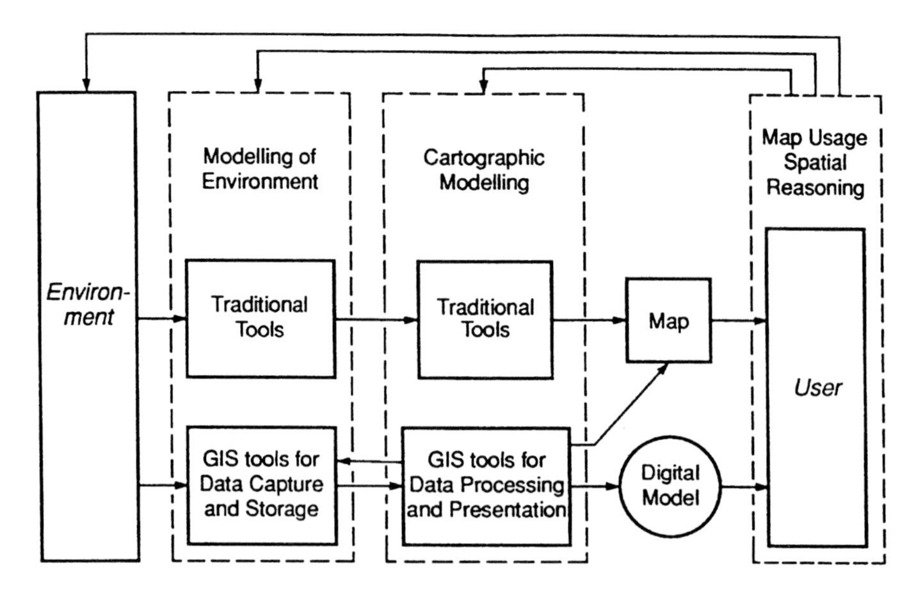

Figure 4.1 Cartographic communication network.

most comprehensive form of spatial models for visual communication and thinking is still the map on screen, paper, etc., and visual data processing is still the fastest and most effective way of information processing that can lead to immediate decisions on human behaviour and action (Freitag, 1993). That is why cartography as a discipline has to play an important role in GIS applications today and in the future.

4.2 Geoinformation systems

4.2.1 Concept

Cartographic model theory has proved to be an appropriate concept for the development and implementation of GIS since the 80s (Grünreich, 1992). The general concept of a GIS is presented in Figure 4.2. Again, it shows the information-oriented relation between *user* and *environment* which is constituted by:

- a **digital object model** (DOM) which comprises
  - a digital object-oriented landscape model (DLM) as a spatial reference system and
  - digital thematic models of all disciplines being integrated into the DOM;

  the DOM can be made up with geodata from different sources, the most useful ones are existing maps;
- **digital cartographic models** (DCM) as the result of purpose-oriented cartographic design processes elaborating on geodata selected from the DOM or on the outcome of GIS analysis;
- **cartographic expressions** (maps, perspective presentations, etc.) both on graphic display or as paper maps which have to be readable and understandable to the users.

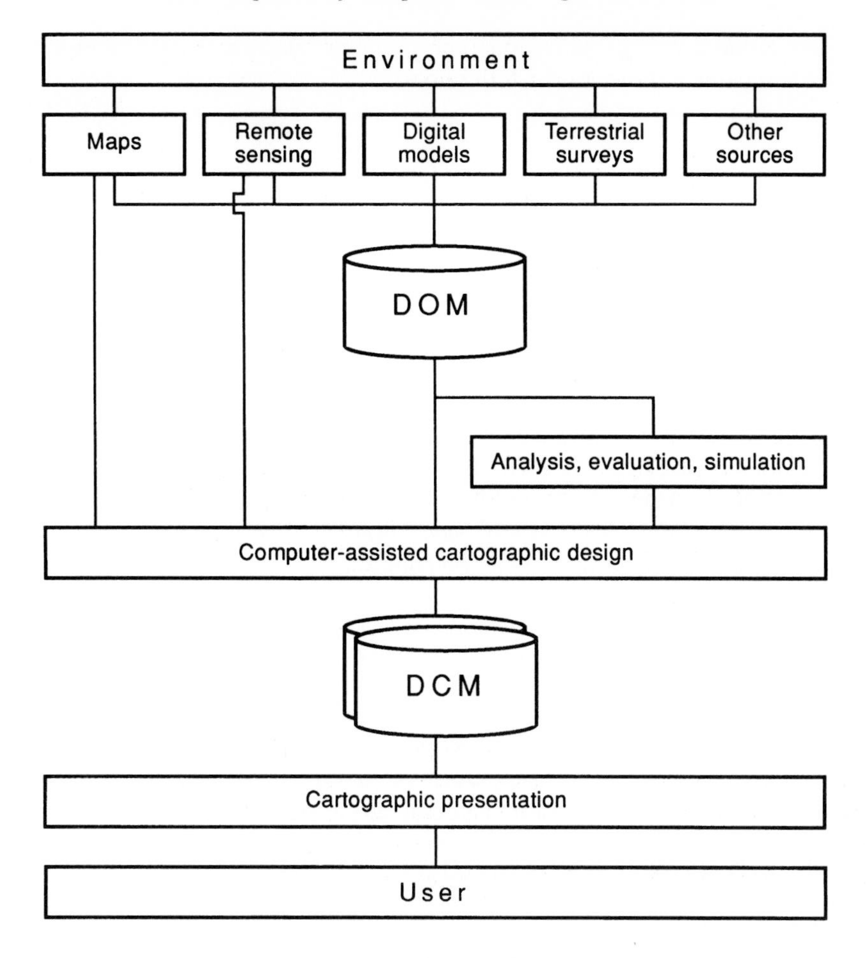

Figure 4.2 Set-up of a geoinformation system.

4.2.2 Implementation of GIS

Digital object models are being established in many European countries and elsewhere. For the following considerations, the topographical GIS of the German state ordnance surveys — the Authoritative Topographic–Cartographic Information system ATKIS — is taken as an example as it was developed on the concept explained above (Grünreich, 1992). Since the beginning of this decade, the German state ordnance surveys have been making strong efforts in order to establish a DLM. This goal can only be reached with a reduced amount of topographical information (compared with the official map series) in the first stage of implementation.

Synchronously, the traditional topo maps are being produced using digital technologies and delivered in the form of raster data in order to meet the requirements of those users who need to work with more topographic information than is provided by the reduced DOM. In other words, societal mapping is both kept on the traditional level of cartographic quality and is going to be extended considerably by DOM. It seems to be reasonable to direct research and development towards a goal which can be described as a combination of DOM usage and a high-quality purpose-oriented cartographic output for the sake of improved spatial information. Computer-assisted generalization of geodata plays a key role in order to reach this goal.

4.3 Generalization in GIS

4.3.1 Task and terminology

Computer-assisted generalization is the main modelling process in a GIS in order to get useful products. There are two kinds of generalization processes: object generalization and cartographic generalization.

1. **Object generalization** is on one hand the process during data capturing in order to set up basic primary models of the environment, and on the other hand to derive primary models of lower semantic and geometric resolution from basic DOMs. The latter is called *model-oriented generalization* or *model generalization*; it is used for numerical applications and also for cartographic generalization.
2. **Cartographic generalization** is necessary for all applications which are based on visual thinking and spatial reasoning. It comprises scale-dependent and design-dependent model generalization and (carto)graphic modelling which is specific for each map.

Figure 4.3 gives a systematic overview on generalization.

A complete and precise description of these processes and their spectrum of application is motivated by the urgent need for *acquiring and formalizing the knowledge* about generalization which is a prerequisite for the development of computer-assisted solutions.

Purpose of G. / Type of Information	Object - G.			Cartographic G.	
	Generalization from Environment to Model		Model - G. (from Model to Model)		
	Environment → A (Base Map)	Environment → D (Base Model)	D → D (Object - M. → Object - M.)	D → D → A (Object - M. → Cartogr. M. → Map)	A → A (Base map → Derived map)
Geometric G.	Influencing details and accuracy of geometric information		Reduction of Geometry of Base model		Reduction of Geometry and Graphic of Base Map
Semantic G.	Qualitative G. : classification				
	according to model resolution		less detailed		graphic dependent
	Quantitative G. : sums, averages etc.				
	according to model resolution		Selection of values		graphic dependent
Temporal G.	Temporal reference of data capturing; dates and time intervals				
	according to model accuracy		less accurate, selective		

Acronyms : G = Generalization, M = Model, A = Analogue Model, D = Digital Model

Figure 4.3 The scheme of generalization of geodata.

Speaking about the content of knowledge it is possible to distinguish between *declarative* and *procedural* knowledge (Meng and Grünreich, 1993). In the context of generalization,

- **declarative** knowledge describes the generalization problem (e.g. how to classify geodata for a certain purpose);
- **procedural** knowledge describes the process of generalization (e.g. what is to be done with given geodata).

4.3.2 Strategy for research and development

Generally, both declarative and procedural knowledge can be represented explicitly (fact, rule) or implicitly (parameter, algorithm). However, to formalize procedural knowledge proved to be far more sophisticated than to formalize declarative knowledge. All known techniques for formalizing knowledge more or less suffer the same drawback that a large amount of knowledge which cannot be verbally described remains neglected. This is also valid for generalization in a GIS environment. Therefore, an evolutionary strategy of research and development is proposed as follows.

1. *Model generalization* should be developed with highest priority (4.3) as it is necessary for numerical GIS applications and for cartographic presentation (4.3.1) as well.
2. The methods developed should then be adapted for *cartographic generalization* starting with standard map series (4.3). At this stage, cartographic modelling such as displacement is to be performed interactively.
3. In order to get the process of cartographic generalization more automated the expertise of cartographers in solving graphical conflicts is to be extracted and formalized thoroughly using their guidelines, style sheets and well-designed maps as standards (4.3).

4.3.3 Research and development of model generalization

The declarative knowledge which is necessary for model generalization can be described as:

- the geo-objects to be derived,
- the geo-objects to be used as input, and
- the quality of object information of input DOM and output DOM, etc.

This kind of knowledge is to be implemented using procedures as classification, amalgamation and simplification. How to process the input model into the output model is a matter of procedural knowledge.

The program CHANGE of the Institute of Cartography of the University of Hannover is an example of a *procedural system* (Grünreich, 1993). Originally, it was developed for cartographic generalization of discrete topographic objects in large scale maps for medium scale topo maps. The results should serve as a draft which has to be edited by a cartographer in order to get the map originals. Most recently, however, this approach also appears to be worthy to develop further for model generalization in the scale range between 1:1 000 and 1:25 000 as it is related to such important fields of application as cadastre, town planning and regional planning, etc.

4.3.4 Research and development in cartographic generalization

Cartographic generalization means increasing the readability of a spatial model by diminishing its semantic information content (model generalization) and by graphically coding the information which is useful for the application of interest, i.e. the pragmatic information content is optimized.

Today, computer-assisted cartographic generalization is not operational. However, *societal mapping* (see Chapter 1) has continuously to be kept at least on the traditional level of quality for the sake of many applications like town planning or geoscientific map construction. In order to fulfil this requirement it appears to be necessary to continue with the production of traditional maps in analogue and digital form, as long as area-covering DOMs and the methods for cartographic generalization are available so that maps of acceptable quality can be produced via DCM (see Figure 4.2).

In order to reach this goal, the author proposes to develop a procedure for integrated processing of object-oriented DOM and scanned up-to-date maps (see Figure 4.4). The

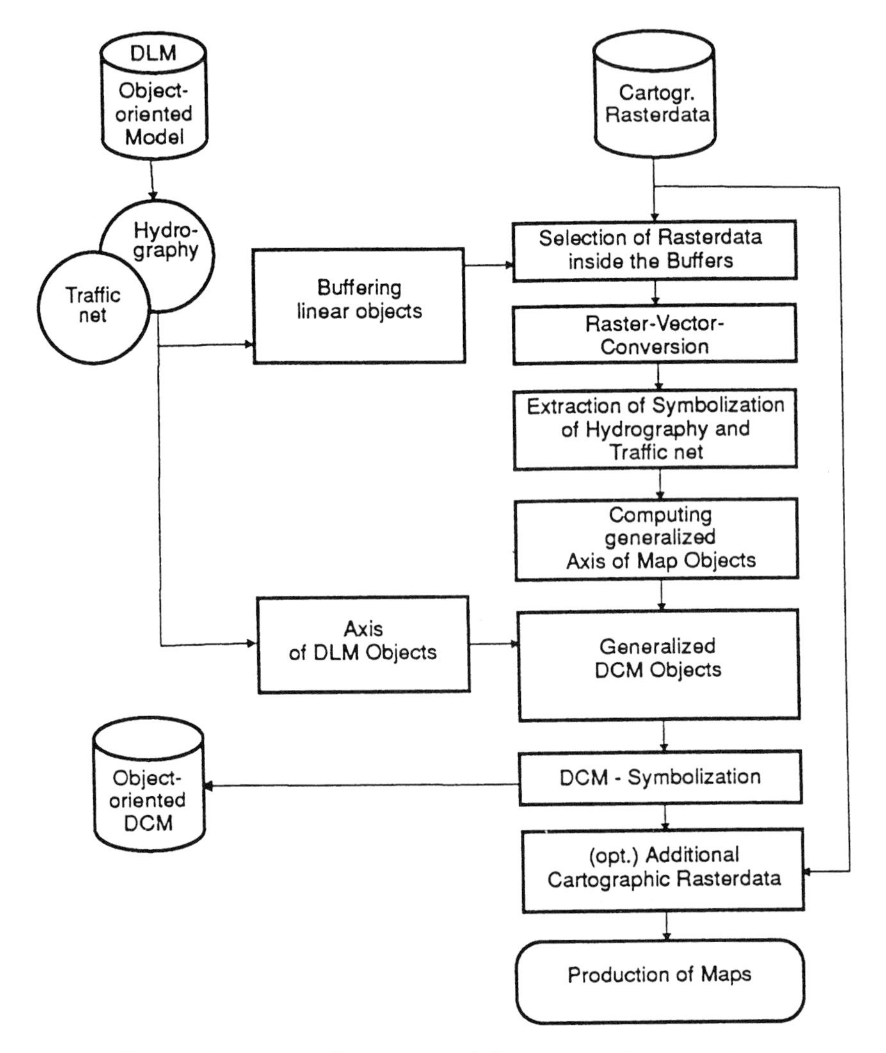

Figure 4.4 Integrated processing of object-oriented and map data.

procedure is characterized by the use of the generalized geometric information of the traditional maps and the object structure of the DOM. Many of the tools to be applied, like buffering, raster-to-vector conversion and computation of the axis of double-lined map symbols, are available in operational status. Some methods have still to be adapted and further developed, especially pattern recognition in order to delete those graphic elements which do not belong to the objects of interest (mainly roads, rivers, etc.), and automatic building of DCM objects by intersecting the axis of the DLM objects and the axis of the (generalized) map objects.

The benefits of this approach are:

1. for national mapping agencies (NMAs)
   (a) the production of traditional maps can be continued using digital technologies but without computer-assisted generalization,
   (b) thematic geodata can be integrated on the basis of digital topographic raster data,
   (c) DOM data can be used to build DCM objects step by step;
2. for research and development
   (a) the procedural knowledge of cartographic modelling can be extracted from a synoptical analysis of a DLM and the according map (see Figure 4.5),

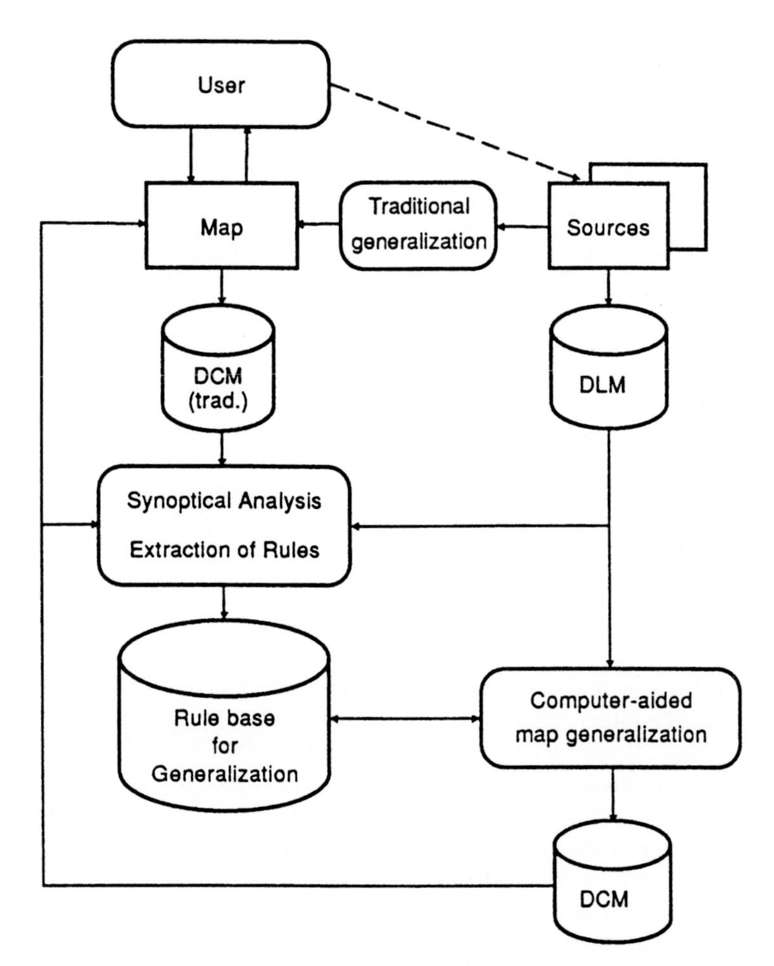

Figure 4.5 Extracting cartographic expertise.

(b) there is more time to get expert procedural knowledge implemented into cartographic expert systems.

4.3.5 Research and development at the Institute of Cartography (IC) of the University of Hannover

In IC's research, the existing procedural system CHANGE is being used as a research tool using the latest GIS technologies. It is planned to establish an object-oriented data structure whose origin can be traced back to the ATKIS feature and symbol catalogue. Both are dealing with *declarative knowledge*, e.g.

- rules for building DCM objects from certain DOM data,
- rules for the graphic design of the map objects, i.e. symbolization, colouring, presentational priority and graphic pattern for each symbol.

In another project it is intended to extract the *procedural knowledge* from a considerable sample of generalized map, the so-called reverse engineering approach (Figure 4.5). The first statistical investigation on the software package CHANGE has been performed in a diploma thesis (Wessel, 1993). By recording the intermediate object states, setting up of parameters, CALL-sequences of procedures and CALL-frequency of each procedure in a log-file during each execution of the software, a statistical foundation for the better organization of the applied procedural knowledge has been laid. Furthermore, the statistical discrepancy between the results of computer-assisted generalization and the traditionally generalized maps has been calculated. Ongoing research has the goal of improving the generalization algorithms, the parameter values for displacement, etc.

4.4 Conclusions

Generalization in GIS is a necessity in order to meet the requirements of societal mapping in information societies, i.e. to provide useful information in an understandable form. Development of appropriate methods should start with *model generalization* at the first stage, because this process is necessary both for spatial analysis and for cartographic visualization. The author is convinced that tools for model generalization can be developed in the near future. Due to the lack of procedural knowledge about cartographic generalization it appears to be worthwhile to continue with generalization performed by human experts (Weibel, 1991) and to merge the results with object models digitally.

In the next step, computer-assisted *cartographic generalization* has to be developed further beyond the interactive editor approach concentrating on methods for resolving graphical conflicts. It appears to be unrealistic that an operational system ready for the production of high-quality cartographic presentations can be developed within the next few years due to a lack of area-covering geodata and sophisticated generalization procedures. In order to provide GIS users (decision makers, regional planners, geoscientists, etc.) with cartographic products of traditional quality, the author proposes to employ digital technologies for updating and the production of paper maps and digital maps (raster data). Digital maps not only serve in many applications but are expected to be valuable sources for extracting explicit knowledge for generalization.

References

Freitag, U., 1987, Do we need a new cartography?, in *Nachrichten aus dem Karten- und Vermessungswesen*, Series II, No. 46, pp. 51–9, Institute of Applied Geodesy, Frankfurt am Main.

Freitag, U., 1993, 'Map functions', invited paper for the public session of the ICA working group on 'Theory in Cartography', Cologne.

Grünreich, D., 1992, ATKIS — A topographic information system as a basis for GIS and digital cartography in Germany, in Vinken, R. (Ed.) *From Digital Map Series in Geosciences to Geo-Information Systems*, pp. 207–16, Federal Institute of Geosciences and Resources, Geologisches Jahrbuch, Reihe A, Heft 122, Hannover.

Grünreich, D., 1993, Generalization in GIS environment, in *Proceedings of the 16th ICA Conference*, Cologne, Vol. 1, pp. 203–10.

Meng, L. and Grünreich, D., 1993, 'A Note on knowledge formalization', paper presented at the 8th NCGIA Initiative Meeting, Buffalo, USA.

Weibel, R., 1991, Amplified intelligence and rule-based systems, in Buttenfield, B.P. and McMaster, R.B. (Eds) *Map Generalization: Making Rules for Knowledge Representation*, pp. 172–86, London: Longman.

Wessel, W., 1993, 'Statistische Untersuchungen bei der automationsgestützten Gebäudegeneralisierung', unpublished Diploma thesis, Institute of Cartography, University of Hannover.

5

Three essential building blocks for automated generalization

Robert Weibel

Department of Geography, University of Zurich, Winterthurerstrasse 190, CH-8057 Zurich, Switzerland

5.1 Introduction

In the past few years, applications of geographic information systems (GIS) have matured, and databases of large size have been built. Users are now beginning to realize the lack of generalization functionality with respect to the development of value-added products from the initial database and the update of existing databases, particularly when multiple scales are involved. Since the majority of results of GIS-based modelling activities is still communicated to the end-user in a graphical form, functions are needed for automated *cartographic generalization.* It must be possible to derive display products from a basic database at arbitrary scale or symbolization, and to maintain good readability. In a digital environment, however, the requirements of generalization extend beyond the original focus on cartography, and include functions for *model generalization* or model-oriented generalization, as will be discussed below.

In order to meet the requirements of today's GIS applications, research needs to tackle various problems. Müller *et al.*, in Chapter 1, have attempted to summarize the state of the art of research in generalization, and outline the problems that would need to be addressed in the future. The discussion here will concentrate on three issues which are considered as particularly important for the progress of generalization research:

- model generalization,
- knowledge acquisition, and
- the evaluation of generalization alternatives.

These elements are considered as 'essential building blocks' for computer-assisted generalization, since they form central prerequisites and seem most needed for a comprehensive solution of generalization in the digital domain. Not all of these problems are on the same level of complexity and functional extent, but they all seem important. The discussion of these three 'pièces de résistance' attempts to identify the subproblems that are involved, and to offer some indications based on our own experience. It is clear, however, that no comprehensive or final solutions can be presented at this point. It is also obvious that these are not the only important issues in generalization. There are other problems that should also receive ample attention, such as the development of more suitable data structures to support map generalization, methods for knowledge representation, or enhanced user interfaces.

5.2 Model generalization

Today, there is a consensus in the research community that, apart from graphics-oriented generalization, there is also a need for model or model-oriented generalization (see Chapter 1).

What does model generalization encompass? One of the major objectives of model generalization certainly is *controlled data reduction* in the spatial, thematic, and/or temporal domain. Data reduction may serve a variety of purposes. A classical aim is to reduce data volume in order to save storage space or speed up computations. Another important reason to reduce the accuracy and resolution of a dataset is the homogenization of different datasets in the process of data integration (or data fusion). For instance, the values of a monthly time-series may need to be reduced to a yearly time interval, in order to develop a unified series. Reclassification or formation of complex objects, which are processes of data reduction in the thematic domain, may be used to prune the solution space for many computations. If data reduction is applied as a filter process (to continuous data), it may be used for the detection and elimination of data errors (Heller, 1990). Besides data reduction, an important objective of model generalization is the *derivation of databases at multiple levels of accuracy and resolution.* This is equivalent to deriving a Digital Landscape Model DLM_2 of reduced contents from an original DLM_1 (Brassel and Weibel, 1988; Müller, 1991a). Finally, of course, model generalization may precede cartographic generalization as a *preprocessing operation.* For instance, the selection of relevant features may be purely model driven and coordinates may be filtered to a resolution corresponding to the target scale of the intended map. This approach corresponds to first reducing a DLM_1 into a DLM_2, and then deriving a Digital Cartographic Model (DCM: Brassel and Weibel, 1988; Müller, 1991a).

It is perhaps interesting to note that, so far, most of the research carried out in model generalization has focused on discrete data, such as objects included in cadastral or topographical maps (Chapter 4; Chapter 18). However, besides discrete objects which can be clearly delineated and discerned, geographic databases also include digital representations of phenomena that vary continuously over space and/or time. Model generalization should, therefore, also include methods to deal with those kinds of data. Examples of continuous variables include terrain, soil salinity, or population density. Commonly, such variables are measured at discrete locations, and the continuous surface is estimated through interpolation. In order to derive a surface of reduced accuracy, the generalization process must start from the original measurements. A possible procedure for model generalization of digital terrain models (DTMs) is demonstrated in Weibel (1992). It is based on an algorithm for iterative filtering of triangulated DTMs developed by Heller (1990). The use of the term 'statistical generalization' by Brassel and Weibel (1988) for what is now commonly known as model generalization was influenced by the work on continuous surfaces reported in Weibel (1992).

In contrast to cartographic generalization, model generalization involves no artistic, intuitive components. Instead, it encompasses probabilistic or even deterministic processes. For the same reasons, engineering-oriented researchers and computer scientists, who frequently had problems understanding the need for artistic compromises in cartographic generalization, should feel more comfortable with model generalization. Thus, one would naturally expect that it should be easier to tackle than its graphics-oriented counterpart. On the other hand, a limiting factor at this time is perhaps that the requirements of model generalization for relevant GIS applications are not yet defined. As in cartographic generalization, different methods will have to be developed for different applications of model generalization. It cannot be expected that 'one size fits all' methods can be devised. It is, however, possible to state some general requirements that should be met by all procedures for model generalization.

- The method should produce predictable and repeatable results.
- The deviations of the resulting model from the original model should be minimized (or at least never exceed a given maximum tolerance).
- The reduction of the data volume should be maximized.
- The integrity (e.g. the topological consistency) of the objects modelled in the original model should not be violated.
- From a user point of view, the procedure should be controllable by as few parameters as possible, and the relation between the input parameters and the result of model generalization should be obvious.
- Finally, efficiency is a further requirement, as model generalization is often aiming at data reduction with the objective of speeding up computations.

Taking the well-known line filtering algorithm described by Douglas and Peucker (1973) as an example, one can observe that it meets the above requirements only partially. It produces predictable and repeatable results, and achieves major reduction factors at little cost in terms of deviations from the original line (McMaster, 1986). Also, it is efficient and can be controlled by a single tolerance value that is easily mappable to the result. The problem, however, is that this algorithm can create self-intersecting lines because no mechanism is included for checking against topological inconsistencies (Müller, 1990). Furthermore, overlaps might result between different lines as a result of filtering each line individually.

The problem with many existing methods for the generalization of spatial data is that they have been developed with no sufficiently focused objective in mind. Although they achieve generalization-like behaviour, it is not always clear whether they can be used for model generalization because they may not fulfil the above requirements. Likewise, many of these algorithms are also not suited for cartographic generalization because they do not pay attention to cartographic principles. The example of the method developed by Douglas and Peucker (1973) is but one of a longer list of possible examples. Nevertheless, these algorithms are used for purposes for which they were never really intended, such as the production of multi-scale databases. It is obvious that future methods for model generalization should adhere more strictly to rigorous criteria that can be used to evaluate their performance, such as the generic requirements outlined above, but also more specific ones.

5.3 Knowledge acquisition

Today, much of the research in cartographic generalization is generally leading in the direction of knowledge-based systems. As a rule, knowledge-based systems derive their power from the knowledge they contain, and not from the particular formalisms and inference schemes they employ. Thus, the formalization of generalization knowledge, known as knowledge acquisition (KA), has become an issue of major importance (Chapter 1). In the computer-science literature (e.g. McGraw and Harbison-Briggs, 1989), the classical methods for acquiring knowledge in the knowledge engineering (KE) process include interviewing experts, learning by being told, and learning by observation. Several supporting techniques have been developed to make knowledge engineering more effective: structured interviews, repertory grids, critical incidents, artificial problems, as well as querying by means of expert system shells. Unlike the primary application domains of today's knowledge-based systems, medical diagnosis, systems configuration, taxonomy, or fault diagnosis, which are based on complex, yet rather well-documented knowledge, generalization involves a great deal of intuition. What makes cartographic knowledge most special, however, is that it is essentially encoded graphically and, thus, hard to describe in words. One rather widely used typology distinguishes between geometric, structural, and procedural knowledge components that are involved in generalization (Chapter 1). Given the specific situation of generalization, it seems natural that the conventional knowledge engineering methods must be refined and extended.

Several different methods are potentially useful for knowledge acquisition in map generalization: conventional knowledge engineering techniques, analysis of text documents, comparison of map series (reverse engineering), machine learning, artificial neural networks, and interactive systems (amplified intelligence). Most previous efforts in knowledge acquisition (e.g. Nickerson, 1991; Mulawa, 1993) have concentrated on the use of conventional KE techniques; that is, they were mainly based on expert interviews. Experience with alternative methods is very limited because the research community is only beginning to tap these new possibilities. On the other hand, it is important to assess the potential of the different approaches and find out about possible strengths and limitations soon. Given the complexity of the generalization process, a combination of KA techniques seems most useful and one would naturally like to direct research efforts to the most promising alternatives.

The author's group are currently conducting studies with several techniques (reverse engineering, amplified intelligence, machine learning, and neural networks) in order to gain some initial experience and structure future research. Based on preliminary results of these studies and on theoretical considerations, the potential of the different alternatives was estimated. A detailed discussion of this effort is given in Weibel (1993). Here, the discussion will concentrate on a summary of the specific characteristics as well as problems that remain to be resolved by future research (see Table 5.1).

As mentioned above, *conventional KE methods*, in particular interviewing techniques, have been used in previous projects. They seem particularly useful for an initial structuring of the problem domain, but also in the long term as a background and complementary strategy for other KA methods. The advantage is that the knowledge is acquired at the source, and thus includes the experts' explanations. However, a fundamental problem encountered in the course of a previous project (Weibel, 1992) is that, because of the holistic nature of the cartographic design process, cartographers find it hard to break down a workflow into distinct actions. Also, they are often reluctant to contribute to technology which they consider is performing clearly below their standards. Similar communication-related problems also exist in knowledge engineering in other fields. Considerable research has thus been carried out on developing communication strategies and interviewing techniques that should help to cope with these problems (McGraw and

Table 5.1 Synopsis of different methods for knowledge acquisition

Method	Time-frame and complexity	Potential use for KA in generalization	Specific characteristics and problems
Conventional KE (interviews, observation of experts at work)	• particularly useful in initial phase, but also long-term • ±low complexity • partial automation possible	• establish initial framework • background for other KA methods • projects in large institutions (NMAs)	• knowledge acquired at the source, includes explanation • needs availability of experts (institutional framework) • experts may be unable or unwilling to explain actions
Analysis of text documents (guidelines)	• only useful during initial phase • low complexity • little automation possible	• initial knowledge base (procedural knowledge) • extensive potential knowledge source	• descriptions often vague • difficult aspects rarely explained in written form • conflicts between rules possible
Analysis of maps: reverse engineering	• only useful during initial phase • ± low complexity • little automation possible	• formalize rules for selection (e.g. Radical Law) • procedural knowledge: semiformal descriptions rather than formal rules	• original generalization idea may be obscured by later updates • final map may not reveal intermediate operations • difficult to determine sequence and relations of operators
Machine learning (ML)	• useful in the mid- to long-term range • high complexity • highly automated	• interpretation of large numbers of facts extracted by reverse engineering or interactive systems • refinement of initial rules	• no previous experience with ML in cartography • so far, lack of suitable data generated by previous investigations • debugging of knowledge necessary

Neural networks (NNs)	• useful in the mid- to long-term range • high complexity • highly automated	• not very useful for KA due to lack of explanation • replacement of algorithmic generalization operators by more holistic approaches • classification (structure recognition)	• very limited experience with NN in cartography • which network topologies work best? • input representation? • choice of training set
Interaction systems (amplified intelligence)	• useful over the entire time-frame • medium to high complexity • automated, but needs human intervention	• evaluation of genealization operators and support facilities (immediate user feedback possible) • KA through interaction logging • integration and testing of knowledge acquired from different sources	• needs involvement of human experts • ‘packaging’ of operators, interaction mechanisms? • limited experience with interaction logging

Harbison-Briggs, 1989). Another problem that is quite typical of cartography is that skilled cartographers (i.e. potential experts) usually work at large institutions. It is thus often difficult, if not impossible, for outsiders to conduct expert interviews — an institutional framework is needed. Therefore, conventional KE methods are often unattractive for academic researchers who traditionally have carried out most of the research in generalization.

The *analysis of text documents*, particularly of compilation and production guidelines in use at mapping institutions, represents another approach that needs little technological investment. Guidelines provide an extensive potential source of semi-formal knowledge, especially of procedural knowledge. The analysis of such documents thus seems particularly useful during the initial phase of knowledge base development. However, the descriptions contained in production guidelines are usually rather vague, incomplete, and particularly fall short of explaining the difficult aspects of cartographic operations (Chapter 17). In some cases, they are even kept mainly in graphical form, showing illustrations of favourable and unfavourable examples (e.g. SSC, 1987). Another problem is that since guidelines are usually written in a 'sequential' fashion, conflicts between different rules may arise.

As an alternative to analysing text documents, graphical documents — maps — may be studied. This approach attempts to extract generalization knowledge by comparison of the modifications that occur to the individual map elements across the scales of a map series. The strategy has also been dubbed 'reverse engineering', since the process starts with the end-product, and attempts to identify the operations that led to this result. In recent years, high expectations have been raised with respect to this strategy (e.g. Buttenfield *et al.*, 1991; Muller 1991b). However, one must be aware of the fact that the final map is usually the product of a series of complex and convoluted design operations. Thus, apart from technical problems involved with measuring and tracing generalization operators, it is frequently impossible to identify reliably the operations that led to the end-product, and determine their sequence and relation. Also, the original generalization idea may be further obscured by later updates if map sheets of a regular map series are used. Thus, in order to portray generalization in an unbiased fashion, the maps used in a study currently conducted in the author's own group are 'new' maps, having been produced specifically for the purpose of this experiment (Parantainen, 1995). Given the difficulties with this method, it appears that while reverse engineering has been capable of extracting quantitative relations such as the Radical Law developed by Töpfer (1974), the usefulness of this method with respect to formalizing procedural knowledge must be seen in a more conservative fashion (Parantainen, 1995). The output of reverse engineering should be considered as semi-formal descriptions rather than formal rules. These descriptions, in turn, may then support the development of more accurate knowledge using other KA techniques. Also, the analysis of maps can often provide a communication link between the knowledge engineer and the expert cartographer.

Machine learning (ML), in the context of knowledge acquisition for generalization, mainly has its merit as an auxiliary technique. Given the prospect of large numbers of facts to be compiled by reverse engineering, or audit trails produced by interactive systems (see below) in the future, some consideration needs to be given to the way in which these observed but unstructured facts are turned into rules. For humans, it is very quickly impossible to perceive patterns in datasets exceeding the size of just a few elements. ML methods (induction, deduction, concept-based learning, statistical clustering, or neural networks) are capable of generating decision trees or prototype rules that facilitate the formulation of an initial rule set from observed facts (McDonald, 1989). ML can potentially also unveil unknown or unexpected relations and rules. However, ML cannot find any rules that are not 'captured' by the original attributes contained in the facts database. Finally, ML may be used to refine initial rule sets. ML techniques (mainly based on induction) are implemented in several commercial expert system shells as well as in public domain packages. The major impediment to the use of ML thus far has been the lack of suitable data to experiment with the application of ML to cartographic problems. For good results, ML techniques need a large number of reliable facts as input. Also, the decision trees or rules generated by ML need to be tested and possibly debugged, since incorrect or conflicting rules may be inferred.

Neural networks (NNs) or more correctly, artificial neural networks, are a specific form of machine learning (Maren *et al.*, 1990). NNs are capable of learning based on training from given sample situations, but it is hard to actually deduct formal knowledge (i.e. rules, decision trees, etc.) from them. With respect to knowledge acquisition, they are thus not very useful. Nevertheless, neural networks are of interest to generalization due to their great ability in classification and template matching. NNs may be useful for several generalization tasks, as is discussed by Werschlein and Weibel (1994). The most straightforward application is the use of NNs for the classification of map feature attributes (i.e. thematic generalization). Another potential use of NNs is in the context of structure recognition, where they could replace statistical methods for the classification of 'structure signatures' (Buttenfield, 1991) with a potentially more robust approach. A third area to which NNs might be applicable in generalization is the evaluation of alternative generalization solutions produced by different methods and/or different parameters (see the next section). Perhaps the most interesting NN application of all, however, is in replacing current algorithmic generalization operators with more holistic solutions. For instance, algorithmic operators for line generalization are split up into subprocesses such as simplification, smoothing, and enhancement, while a suitable caricature can often only be obtained through a combination of processes. NNs may potentially overcome this discretization.

As with machine learning, however, practically no previous experience exists with the use of NNs in cartography. Initial work with NNs in line generalization performed by Werschlein and Weibel (1994) suggests that the performance of NNs primarily depends on three points: the scheme that is used to represent the input data, the topology of the network, and the choice of samples used to train the network. Among these factors, the choice of input representation is of overriding importance. The representation of the input data essentially dictates what patterns can be inferred by the neural network. In raster-based generalization (e.g. of land use maps), input representations can be restricted to simple raster data structures, since generalization is basically performed by reclassification of cells, while geometry does not change. In vector mode generalization, however, input representations cannot be restricted to the basic data structures used by algorithmic methods (e.g. simple strings of x/y coordinates). Basic representations must be enriched by additional transformations (e.g. line curvature) in order for the neural network to be able to infer reliably shape modifications between input and output. Werschlein and Weibel (1994) provide a discussion of possible input representations for cartographic lines.

The last alternative for knowledge acquisition is the use of *interactive generalization systems*, also termed *amplified intelligence* (Weibel, 1991). The basic idea is that interactive systems could be equipped with a facility for logging the interactions of expert users with the system. The analysis of the resulting interaction logs (also called audit trails) is then expected to lead to the formulation of rules. This author proposed this approach several years ago (Weibel, 1989, 1991); other researchers are now pursuing similar strategies (e.g. McMaster and Mark, 1991). The concept is appealing, even more so since the same set-up could also be used for testing of existing and new generalization operators and support facilities, as well as the integration and testing of knowledge that is acquired through different methods. On the other hand, systems that carry some potential to act as a platform for knowledge acquisition are only beginning to emerge (e.g. Lee, 1993), and a number of difficult problems still need to be resolved before this approach can be exploited successfully. Formats for interaction logs (what actions are being logged, when, and how?) and mechanisms for editing these logs must be developed. The 'packaging' of operators — that is, the level at which the user can control them — will significantly influence the interaction logs that can be produced. Furthermore, the characteristics of map features must be determined and logged as well (e.g. line sinuosity, relation to neighbouring features, etc.), in order to identify why generalization operators have been applied and what triggered these actions. Once interactive systems are being used for interaction logging experiments, they are bound to generate a vast number of facts describing the flow of operations during generalization. It will therefore be necessary to use machine learning tools in an attempt to interpret these unstructured facts. One expects, however, that the automated interpretation of interaction logs alone will not suffice; further interviews and session observations (possibly by video taping) of the experts at work are needed as complementary techniques. Finally, it should be mentioned that audit trails can have more direct uses besides knowledge acquisition, which can also be beneficial in a production environment. The most immediate use of interaction logs can be seen in the creation of macros for action replay. Audit trails may also be used for 'generalization by example': parameters are interactively trained and logged for one or several smaller but representative regions of the original map, and subsequently applied to the entire map automatically.

In conclusion of this section, one can observe that a variety of methods can be explored for KA in generalization that have the potential of complementing each other in terms of the types of knowledge that may be formalized, the technological efforts involved, the degree to which experts are involved in the KA process, and their current maturity. Conventional KE techniques and analysis of guidelines seem most valuable in the context of research conducted by or in close collaboration with mapping agencies. The position of reverse engineering should be seen conservatively; its best use may be as a means of bringing together researchers from the academic sector and practising cartographers. Machine learning and neural networks represent novel techniques in cartography that should be studied extensively by academic research. Interactive systems offer the best potential for the integration and testing of knowledge from various sources and should therefore be pursued even if future research shows that their value for knowledge acquisition is limited.

5.4 *Evaluation of generalization alternatives*

While some research was conducted in the late 1970s and early 1980s to develop geometric measures for an assessment of line simplification (e.g. McMaster, 1986), the development of methods for evaluating generalization results has received very little attention since then. It is only now being realized that such evaluation methods are an important component and even a prerequisite of knowledge acquisition. Evaluation is needed at three different stages of the generalization process.

- *A priori evaluation* is necessary prior to the actual knowledge acquisition process. It includes the selection of suitable study areas and test datasets, and helps in assessing the performance of potential expert cartographers as well as distinguishing examples of good solutions from poor work which should not be considered for knowledge formalization. The latter aspect is particularly crucial for the selection of suitable training samples for NN applications. Often, some of the input materials may still be in analogue form and may need to be digitized first.
- *A posteriori evaluation* is needed to compare and rank different generalization alternatives: for instance, those that are generated by different generalization operators or sequences. It also involves the assessment of existing algorithms and techniques for generalization (both generalization operators and support facilities). In principle, the methods that can be used for a posteriori evaluation are the same as those for a priori evaluation.
- *Ad hoc evaluation* is required to control the automated generalization process as a means of continuous evaluation. It includes the task of conflict resolution between contradicting rules at run time and thus forms part of the meta-knowledge of a knowledge-based system. Some approaches such as the so-called genetic algorithms involve the automated generation of large sets of alternatives using different parameters in order to find the best solution (Armstrong, 1993). Ad hoc evaluation is needed in this case to prune the solution space and to determine the optimal solution.

The overall process of evaluation involves two tasks. First, *specifications* must be established of what defines a 'good' or 'acceptable' generalization for a given map. It is only against such requirements that a generalization result can be assessed. These specifications, of course, will depend on the constraints of the generalization process, such as the purpose and scale range of the given maps, and the quality of the input data. Apart from criteria relating to the quality of the graphic output, other factors such as efficiency, robustness, and ease of use are relevant when evaluating digital methods. As a second task of evaluation, the actual *assessment* must then attempt to determine the degree to which the specifications are met by a given result.

The study of the traditional cartographic literature can provide a natural starting point for the development of methods for evaluating generalization alternatives. Many textbooks and teaching notes (e.g. SSC, 1987) make use of 'good' and 'poor' examples to explain particular concepts of generalization. An investigation of that literature can provide some initial hints for the development of specifications as well as methods for assessment. Further input can be gained by studying the quality assurance procedures currently in place at mapping agencies for manual production. And finally, a study of the literature on quality assurance and benchmarking in non-cartographic disciplines such as industrial or software engineering could perhaps furnish further insight about methods for the assessment of non-quantifiable processes.

Since generalization involves highly intuitive components, it cannot be expected that any evaluation procedure can be purely objective. Definitely, there are criteria that can be assessed objectively such as violations of topological relations, but other aspects such as 'overall clarity of the map image' are very hard to express rigorously. These objective and subjective components of generalization can be expressed by a number of quantitative and qualitative criteria. In general, quantitative measures relate to objective properties of generalization and design, while qualitative descriptions are used to capture subjective aspects. However, note that 'quantitative' and 'objective' are not synonymous: minimal dimensions, for instance, can be specified in a strictly quantitative way, but the actual values used for a particular map are often left to specific subjective preferences. Likewise, qualitative descriptions of subjective components can also be made in a more quantitative fashion, for instance, by using a grading scheme and associated weighting factors. Both quantitative and qualitative criteria must first of all allow us to compare different generalization alternatives (among themselves and/or against a solution that is considered optimal), and eventually make it possible to judge and rank different solutions consistently. The question is then how objective and subjective aspects of generalization are characterized in a meaningful way: at which stage of the evaluation and in which way should quantitative measures form the basis of a qualitative assessment?

A project currently under way in the author's group is the development of a methodology for the evaluation of generalization alternatives (Ehrliholzer, 1995). This includes the definition of a prototype report format that integrates quantitative and qualitative criteria, as well as sets (or modules) of measures and qualitative aspects that are considered relevant for different types of generalization problems (depending on map purpose, scales, feature classes, and data types). The expectation is that, in a particular evaluation situation, a specific report format could be compiled from relevant modules, and the resulting values for the report items weighted according to the preferences defined for the project.

Quantitative measures have the purpose of supporting the assessment by determining computationally to what degree design specifications are violated and/or how closely an 'optimal' solution (e.g. a digitized template of a manually produced version) is matched. Possible measures include the following:

- **Global measures**. These may include the degree of generalization and whether it is constant over the entire map (measured by means of feature density and feature clustering), adherence to the Radical Law, and the ratio of foreground to background (b/w ratio).
- **Geometrical measures**. A first group of geometrical measures is needed to highlight cases where the *minimal dimensions* are violated (i.e. objects that are too small, objects that are too close, segments that are too short, etc.). A second group is required to determine the amount of *distortion*, that is, deviation from the *shape* of the representation in the original map or in a map that is used as a reference (e.g. a manually produced map that is considered as a good solution). McMaster (1986) describes a number of such measures for linear objects. Similar measures need to be developed for point and area objects. Various measures are readily available for this purpose from the literature in the fields of computer vision as well as geography and cartography (e.g. Pavlidis, 1978; Austin, 1984, to name but two possible sources).
- **Topological measures**. The purpose of these measures is to identify violations of topological relations that need to be maintained from the original map. These include the detection of self-intersections of lines as well as intersections among different lines, overlapping objects, and adjacency relations that are violated (e.g. a house moved to the other side of a road as a result of simplifying the road).
- **Software-related measures**. In order to measure the productivity gain that can be accomplished as a result of automation, several aspects of software performance must be assessed, including CPU time, person hours spent on a particular generalization task, equipment hours and cost, entire duration of the project, and possible map update cycle. The person hours, to some degree, reflect the ease of use and the robustness of the generalization software. These aspects, along with other more

evasive software-related factors, such as work satisfaction, must be assessed further in the qualitative part of the evaluation procedure.

The implementation of a rigorous procedure for *qualitative evaluation* is inherently more difficult to achieve than the formulation of quantitative measures. Apart from the study of the relevant literature mentioned above, a close collaboration with different experts is necessary to develop a suitable checklist and questionnaire that will eventually lead to a common form of assessment. Ideally, it should be possible to compare the evaluations of different generalization alternatives even if they are performed by different experts.

The author believes that the format of a standardized questionnaire and checklist that asks the evaluating expert to give grades (e.g. 1 to 5) for a variety of assessment items is the most effective way to characterize rigorously and consistently subjective aspects of generalization, and also provides the most direct form of integration with quantitative measurements. In order to allow integration, qualitative questions must match the categories of the quantitative assessment. For instance, as a counterpart of global quantitative measures, qualitative aspects on the global level, such as 'maintenance of the overall character of the original map', can be assessed. Similar questions can be formulated as counterparts of geometrical, topological, and software-related measures, where factors such as 'ease of use' are largely subjective.

Of course, in some cases, an expert will find it hard to assign a grade which can only reflect an average for a particular assessment question. It must be possible for him/her also to document specific problems, typical situations, or details that are handled particularly well. A possible solution is the use of hypermedia techniques to implement an evaluation report, allowing integration of screen snapshots, annotations, and sketches as a means of illustrating specific points. The question is then also whether the assessment should be performed completely in the digital domain (e.g. by means of a visual overlay of scanned manual maps and digital solutions). Finally, it should be noted that the distinction of the effects of generalization from those of other cartographic processes may pose a serious problem. For instance, it may be that, in principle, the elements of a particular map have been generalized adequately, yet an inappropriate symbolization of the individual elements may negatively affect the clarity of the map (Baumgartner, 1990). In such a case, it may be very hard to determine what mistakes have contributed to the poor appearance of the resulting map.

5.5 Conclusions

It appears that research in generalization has entered a new stage. In contrast to earlier years, when research in generalization usually concentrated on the development of narrow, special-purpose solutions, current research strategies are attempting to approach the overall problem in a more comprehensive way. Strategies such as amplified intelligence are pursued that can integrate existing methods to exploit their combined potential. Generalization is no longer viewed as a purely graphics-oriented process, but consideration is also given to reduction, reclassification, and filtering processes in the numerical domain, that is, to model generalization. Finally, attempts are being made to base generalization on a better understanding of the processes that are involved, leading towards the development of knowledge-based approaches.

As is often the case when a new stage is reached in a research domain, there are more questions than answers. At this point, it is first of all important to ask the right questions, and develop a concept about where the answers could be found most probably. Based on the example of three issues that we have identified as crucial for progress of research — model generalization, knowledge acquisition, and the evaluation of generalization alternatives — we have attempted to identify problems and potential solutions for research.

References

Armstrong, M.P., 1993, A coarse-grained asynchronous parallel approach to the generation and evaluation of map generalization alternatives, in *Proceedings of the NCGIA Specialist Meeting I-8 'Formalizing Cartographic Knowledge'*, pp. 37–43.

Austin, R.F., 1984, Measuring and comparing two-dimensional shapes, in Gaile, G.L. and Willmott, C.J. (Eds) *Spatial Statistics and Models*, Dordrecht: D. Reidel.

Baumgartner, U., 1990, *Generalisierung topographischer Karten*, Cartographic Publication Series, Vol. 10, Zurich: Swiss Society of Cartography.

Brassel, K.E. and Weibel, R., 1988, A review and framework of automated map generalization, *International Journal of Geographical Information Systems*, **2**(3), 229–44.

Buttenfield, B.P., 1991, A rule for describing line feature geometry, in Buttenfield, B.P. and McMaster, R.B. (Eds) *Map Generalization: Making Rules for Knowledge Representation*, pp. 150–71, London: Longman.

Buttenfield, B.P., Weber, C.R., Leitner, M., *et al.*, 1991, How does a cartographic object behave? Computer inventory of topographic maps, in *Proceedings GIS/LIS '91*, Vol. 2, pp. 15–104.

Douglas, D.H. and Peucker, T.K., 1973, Algorithms for the reduction of the number of points required to represent a digitized line or its caricature, *The Canadian Cartographer*, **10**(2), 112–23.

Ehrliholzer, R., 1995, 'Development of Methods for the Evaluation of Generalization Alternatives', unpublished MSc thesis, Department of Geography, University of Zurich.

Heller, M., 1990, Triangulation algorithms for adaptive terrain modeling, in *Proceedings of the Fourth International Symposium on Spatial Data Handling*, Zurich, Switzerland, 23–27 July 1990, Vol. 1, pp. 163–74.

Lee, D., 1993, From master database to multiple cartographic representations, in *Proceedings of the 16th International Cartographic Conference*, Cologne, Vol. 2, pp. 1075–85.

Maren, A.J., Harston, C.T. and Pap, R., 1990, *Handbook of Neural Computing Applications*, San Diego: Academic Press.

McDonald, C., 1989, Machine learning: a survey of current techniques, *Artificial Intelligence Review*, **3**, 243–80.

McGraw, K.L. and Harbison-Briggs, K., 1989, *Knowledge Acquisition: Principles and Guidelines*, Englewood Cliffs, NJ: Prentice Hall.

McMaster, R.B., 1986, A statistical analysis of mathematical measures of linear simplifcation, *The American Cartographer*, **13**(2), 103–16.

McMaster, R.B. and Mark, D.M., 1991, The design of a graphical user interface for knowledge acquisition in cartographic generalization, in *Proceedings GIS/LIS '91*, Vol. 1, pp. 311–20.

Mulawa, L.I., 1993, Knowledge base system technology in the Defense Mapping Agency's Digital Production System, in *Proceedings of the NCGIA Specialist Meeting I-8 'Formalizing Cartographic Knowledge'*, pp. 165–81.

Müller, J.-C., 1990, The removal of spatial conflicts in line generalization, *Cartography and Geographic Information Systems*, **17**(2), 141–9.

Müller, J.-C., 1991a, Generalization of spatial data bases, in Maguire, D.J., Goodchild, M.F. and Rhind, D.W. (Eds) *Geographical Information Systems: Principles and Applications*, Vol. 1, pp. 457–75, London: Longman.

Müller, J.-C., 1991b, Building knowledge tanks for rule based generalization, in *Proceedings of the 15th Conference of the International Cartographic Association*, Bournemouth, pp. 257–66.

Nickerson, B.G., 1991, Knowledge engineering for generalization, in Buttenfield, B.P. and McMaster, R.B. (Eds) *Map Generalization: Making Rules for Knowledge Representation*, pp. 40–56, London: Longman.

Parantainen, L., 1995, 'Knowledge Acquisition by Comparison of Map Series: The Generalization of Forest Parcels on the Swiss National Map Series', unpublished MSc thesis, Department of Geography, University of Zurich.

Pavlidis, T., 1978, A review of algorithms for shape analysis, *Computer Graphics and Image Processing*, **7**, 55–74.

SSC (Swiss Society of Cartography), 1987, Cartographic generalization, 2nd Edn, Cartographic Publication Series, Vol. 2, Zurich: Swiss Society of Cartography.

Töpfer, F., 1974, *Kartographische Generalisierung*, Gotha: VEB Hermann Haack.

Weibel, R., 1989, 'Konzepte und Experimente zur Automatisierung der Reliefgeneralisierung', unpublished PhD dissertation, Department of Geography, University of Zurich, 264 pp.

Weibel, R., 1991, Amplified intelligence and rule-based systems, in Buttenfield, B.P. and McMaster, R.B. (Eds) *Map Generalization: Making Rules for Knowledge Representation*, pp. 172–86, London: Longman.

Weibel, R., 1992, Models and experiments for adaptive computer-assisted terrain generalization, *Cartography and Geographic Information Systems*, **19**(3), 133–53.

Weibel, R., 1993, Knowledge acquisition for map generalization: methods and prospects, in *Proceedings of the NCGIA Specialist Meeting I-8 'Formalizing Cartographic Knowledge'*, pp. 223–32.

Werschlein, Th. and Weibel, R., 1994, Use of neural networks in line generalization, in *Proceedings of EGIS '94*, Paris, 30 March–1 April, pp. 76–85.

SECTION III

Object-Oriented and Knowledge-Based Modelling

6

Data and knowledge modelling for generalization

A. Ruas and J.P. Lagrange

Service de la Recherche-IGN, 2 Avenue Pasteur, 94160 Saint-Mandé, France

Foreword — Generalization: precision variation or change of perception level?

There is a classical tendency to think that generalization consists of selecting a source of information with a given precision factor and in simplifying its geometry in order to represent it with less precision. Thus, generalization should be essentially a geometric computation process consisting in analysing and simplifying geometric characteristics of features as well as their spatial relations. Information needed to generalize is supposed to be contained in initial data and completed by a visual recognition of geometric characteristics and spatial relations. This assumption may hold in the case of manual or interactive digital generalization. The underlying hypothesis is that there is a sufficient amount of information in the initial database to carry out only one possible data generalization. As a matter of fact, when a map or a database is being produced, only one type of information is chosen. Obviously there is no such thing as being exhaustive, even in a topographic map which is basically supposed to represent distinguishable geographic features. A topographic map is a clever compromise between a map of relief, land use, names, settlements, hydrography and communication networks (Cheylan, 1989). A map is a view of the world, and basic maps produced by National Mapping Agencies (NMAs) show this compromise corresponding to a common need. A map corresponds to a particular view of a geographic space (Piron, 1993). The perception of geographic space changes with the change in scale. When the scale varies, space can no longer be depicted in the same way, by using the same objects. Conversely, a change in perception induces a change in the scale level. So one may wonder if:

- the initial database contains enough information to allow generalization?
- there are one or many kinds of generalizations?

It may be assumed that the difficulty and the cost of generalization caused NMAs to put forward only one type of generalized map for a given scale. However, if a flexible and effective means of generalization existed, it would certainly incite them to carry out different kinds of generalizations according to the point of view of the geographic information. Lanza and La Barbera (1993) proved, in this respect, that a water network can be generalized differently, according to the criteria of hierarchy utilized.

This study is mainly confined to the topographic aspect and it focuses on modelling information appropriate or necessary for generalization. But it is certainly vital to increasingly integrate more flexible generalization criteria to represent geographic space in different ways in order to meet all the demands that users may have. As David Mark emphasized:

> Basic research will be needed to test the adequacy of the cartographic data model for representing all aspects and concepts of geographic space (Mark, 1992).

Generalization should be seen as a process allowing us to perform a change in the perception level of geographic data. Precision and geometry changes are no more than consequences of this process. The authors do not claim that definite answers will be found in this paper, rather we have tried to point out issues that seem to be of prior importance.

6.1 Knowledge needed for generalization

6.1.1 Action versus modelling?

When a cartographer generalizes, his/her objective is to maximize the quality of the information preserved: surely, the definition of quality is subjective as the conflicts in space found in the process of generalization will oblige him/her to make choices (e.g. Should spatial localization of features or their shape be a priority? Is it worthwhile to keep a diversity of information and to reduce the number of features or to select only certain themes?). These choices depend on the specification of the product but also on the final scale and the scale reduction factor. When the scale decreases (let us say under 1:15 000), the preserved features have to be significantly enlarged in order to make them visible. Numerous spatial conflicts appear and they are essentially resolved by: a reduction in the number of features, a simplification in shape, or moving features around in the densest areas.

The lower the scale, the falser is the location of features and the more their shape is simplified. In other words, computed information is an abstraction of the initial data. The main issue of generalization is to find the characteristics allowing us to abstract the data. It is therefore clear that generalization cannot be performed without a preliminary analysis of semantic and spatial relations of features as well as their geometric characteristics. Every generalization operation should take into account relations and properties whether as a means or as a constraint. This leads to the fundamental question stated in Bjorke and Midtboe (1993): What is information and how can information be measured?

6.1.2 What is to be thought of a rule-based system?

A cartographer uses knowledge. It naturally comes out that a formalization of this knowledge by means of rules (either production rules or Horn clauses) should be convenient for automating the generalization of all types of data. Unfortunately, to date, such attempts have not been satisfactory (Weibel, 1991; Chapter 1, this volume), the main reasons for this failure being among others:

- the knowledge actually used by the cartographer turns out to be difficult to identify, so the rules expressed are either too general (e.g. if a feature is too small, then it can be

omitted) or too specific (e.g. if a lake is too small, then it should be omitted, except if it is part of a set of small lakes close to each other, in which case . . .);
- the mechanism of analogy is often used but it is difficult to integrate into a purely deductive system.

Therefore, rules are not generic enough as:

- they often happen to contain a mixture of topology and semantic notions (e.g. 'connected roads should remain such');
- numerous rules have exceptions which also have their own exceptions. Deep knowledge (Chapter 1) is missing. The inference engine may lead to incoherent results.

As a result, should the idea of formalizing geographic knowledge be discarded in favour of an interactive generalization which is a sort of manual generalization? It seems rather that the present failure is partly due to the fact that the knowledge formalized is ill-adapted to generalization, partly because of a lack of means of evaluating conflicts and identifying spatial relations between objects and geometric properties of objects.

6.1.3 What knowledge is used by the cartographer?

When digital data are used, features and even information describing them tend to be regarded independently. This is due to the fact that only a sequence of points and an identification code are needed to represent features graphically. Interpretation of geometric properties and spatial relations between features can easily be made visually and informally. Thus, these relations do not need to be coded. On the contrary, when generalization is being carried out, the following questions will be posed.

- Insofar as the analysis level is modified, do objects remain the same?
- How can the characteristics of an object or a set of objects be preserved?
- Are the distribution laws of objects identical for two perception levels?

Assuming that basic semantic objects are called objects, it is tempting to express most generic knowledge that the cartographer uses for generalization in the following way.

- If an object is suitable for the targeted analysis level, it should be preserved or created from other objects. The notion of suitability depends on the objectives of the map or on the database to be created, as well as on the role of the object in relation to its neighbourhood. An object may be a representative or a synthesis of a set of objects. It can also be suitable because it acts as a link between two important objects (e.g. a secondary road kept solely because it connects a town with a tourist site).
- The geometry of objects preserved or created should correspond to a new spatial resolution. The preserved objects will have a simplified geometry with the exception of characteristic shapes which will have to be kept or amplified.
- If objects have a mutually important relative position, this relative position should be preserved (e.g. connex objects should remain connex, close objects should remain close).
- The result of every cartographic work should be legible. Thickening due to symbolization, overlapping of neighbouring objects, too great a density of information, too small or too narrow objects, should be avoided.

Therefore, it is indispensable to preserve in the best possible way geometric properties, and spatial and semantic relations in the process of generalization while respecting graphic limitations which depend on symbolization and the new spatial resolution. Thus, if generalization is to be automated it is necessary to satisfy the following conditions.

- Identify properties and relations between objects needed for generalization. These properties and relations include geometric characteristics, topologic and spatial relations and semantic properties. This process has been termed 'Structure Recognition' by Brassel and Weibel (1988).
- Find the best possible formulation and representation of properties.
- Define the means of finding those properties in a set of data.
- Be able to use them in a generalization operation.

6.1.4 What meaning is to be given to complex objects?

The definition of a complex object is not quite clear. It may be considered as a geographic object composed of elementary geographic objects. In a classic Geographic Information System (GIS) comprising the notion of complex objects, it can be seen that, most frequently, a complex object is a combination of simple objects of the same type which, instead of being labelled with one common attribute value, reference a single complex object. Complex objects simplify some queries (e.g. 'spotting road number 7'), but in the process of a data exchange, they are often disposed of and recomputed, if necessary. These objects do not supply any further information to the database, they just contribute to localize information.

Other types of 'static' complex objects showing a combination of simple objects of different types are actually rather rare. They may be considered as the first step towards a multiple representation of data or simply towards a creation of different levels of perception in a classical database.

In the process of generalization, geographic data from one or many databases is used in order to create a new database with its own data schema. Certain objects will be created from a set of objects that will no longer exist. Those objects which could have been considered as complex ones in terms of their ungeneralized database are regarded as elementary objects for the new perception level. In fact, classical complex objects become simple objects during generalization. The only real complex objects that will be required are ephemeral ones and they will be used to describe distribution or geometric characteristics of a set of simple objects (e.g. in establishing a hierarchy of a water network, in configurations of a road network in an urban area, in types of geometric lines, etc.). These 'complex objects' will be defined in section 6.3 by labelling them with more precise names (i.e. geometric objects, structural objects, etc.).

6.2 *Information needs and modelling for generalization*

6.2.1 Geometric properties of objects

Generalization needs

Numerous algorithms for geometric simplification exist and they are basically filtering and/or smoothing algorithms. The problem now consists in the choice of an algorithm and the parameter values which depend on the objects. Since the geometry of objects is not uniform, knowledge of geometric characteristics of each object should render the choice of an algorithm and of its parameter values more sensible (Plazanet *et al.*, 1994).

Frequently, a bandwidth approach is used to decide whether a point of a line should be kept; this method does not take into account the fact that the line is in fact a shape (or is composed of shapes). Numerous experiments have shown that this criterium is insufficient. This was the object of the study by Buttenfield (1991) in which the author emphasizes the importance of analysing the shape of an object before generalizing it: 'Knowledge-based simplification requires that the amount and type of detail in the digital file are defined before the algorithm begins to operate, and that expectations of the amount and type of details that should be retained or eliminated at the reduced scale are also defined'. Identification of characteristic shapes is necessary on different levels of perception. On the whole, one should be able to combine points which are part of a geometrically homogeneous section; then, more locally, it may be necessary to describe the geometry of a characteristic shape. For example, one may have to divide a road into sections with a homogeneous geometry (sinuous sections/straight lines) and then describe each bend of each sinuous section. Besides, it is worth noting that shape maintenance, and the measurements performed in order to assess it, are part of the indispensable evaluation of the generalization result quality.

Description elements

Firstly, it is simpler to describe line features. The essential geometry of a surface feature will thus be grasped by the characteristics of its outline and its skeleton. Further information can be included, such as the identification of elongated or ovoid shapes.

To describe a linear object it should be known whether it is: composed of straight lines, of regular curves or of accidental shapes, composed of characteristic, regularly repeated shapes, or sinuous with a fractal tendency. Concerning its characteristic shapes, do they have: a regular frequency and bandwidth, a prevailing direction, acute or right angles, or cusps or loops according to the prevailing direction of the object?

It may turn out that the vocabulary used to describe a line is often inaccurate and sometimes repetitive. There is a marked tendency to describe the geometry of objects by using a vocabulary related to the nature of the objects represented, e.g. mountain roads, meandering water-streams; this often gives a more precise idea of the general shape but very little spatially localized information.

Geometric modelling

The simplest geometric modelling consists of representing lines by sequences of arranged coordinates $\{(x,y)\}$. A line can also be described by encoding changes in direction (i.e. Freeman encoding chains (Freeman, 1978)) or polar coordinates. These descriptions are insufficient as they do not allow identification of object shapes. The difficulty of finding a better formulation is due to the fact that most topographic objects have a complex or even transcendental shape. Geometric modelling should allow identification of the geometry of a line in the following way:

- globally, in order to use the appropriate algorithm to modify a line as a whole;
- locally, in order to enable the identification of a characteristic shape and its local modification.

Global description of a line

Whether the global description of a line is qualitative or quantitative, analogical or analytical, it may be assumed that identification of characteristics will be made through mathematical means, most of which are relatively simple, such as computations of angles, distances or surfaces. However, numerous problems remain to be solved:

- an identification of line types that allows a complete and non-redundant taxonomy;
- criteria revealing line description;
- the interpretation of classical geometric characteristics (angles, distances, etc.);
- line-splitting methods necessary to define fragments homogeneous in geometry. The Douglas and Peucker algorithm (Douglas and Peucker, 1973) presents obvious limitations in this respect.

The definition of both taxonomy and criteria should be attempted to enable the classification of every line type and the qualification of each line. The solution could consist of defining cartographies of line types according to classical criteria of geometric measurements while taking into account acceptable value variations and problems of similarity. The authors' approach is to investigate possible characterizations and geometric representations and, in parallel, to evaluate classical algorithms on typical feature classes. This point is not developed here, the focus being put on modelling for generalization.

Description and characterization of shapes

The description of local geometry can be based on the identification of characteristic points and a simple mathematical formulation of the geometry of the line between two typical consecutive points. Different analyses have been proposed in the past. They may be classified as follows.

- Techniques which rely on direction changes: Freeman chain encoding (Freeman, 1978) and 'energy' measurements (Williams and Shah, 1992; Bjorke and Midtboe, 1993). The main difficulties in these approaches are to manage problems due to the fact that points are not equidistant, and to divide space into a discrete and significant number of direction zones.
- Fractal analysis: 'The notion of fractal geometry is closely linked with invariance properties through a change in scale' (Gouyet, 1992). Various works have been produced to study the fractal dimension of a line (Dutton, 1981; Buttenfield, 1985; Müller, 1986). It turns out that certain types of natural lines — particularly coastlines — do have a fractal tendency. Such characteristics of the line length are parameters that can be used to qualify a line, or to evaluate results of a smoothing.
- Attempts for global characterization: (Buttenfield, 1985; McMaster, 1987; Buttenfield, 1991; Chapter 12, this volume) have defined criteria for characterizing a line feature or defining its structure signature (for example using strip trees (Ballard, 1981)). The usefulness of these parameters has been proved by the fact that they have enabled the identification of differences in the structure signatures of lines chosen according to intuitive criteria of geometrical homogeneity.
- Detection of characteristic points. Different studies have shown that the identification of characteristic points of a line seems to follow rather objective rules as a varied human population, cartographers and others, select the same points, corresponding to the inflexion points and to the vertices. As a matter of fact, the success of Douglas and Peucker's filtering algorithm may be explained by the fact that the points it selects approximate the line vertices quite well. However, line vertices are not sufficient for characterizing a line. Numerous studies have been made on the utility of the inflexion points and on the different techniques allowing identification of them (e.g. Freeman, 1978; Thapa, 1989; Affholder, 1993).

Work accomplished in the field of artificial vision, which often focuses on curvature extrema (Hoffman and Richards, 1982; Leyton, 1988; Lowe, 1988; Milios, 1989; Rosin and West, 1990; Mokhtarian and Mackworth, 1992; Rosin, 1993a, b), could also provide interesting input for this research axis. We do not consider relief characterization and classification here although it is of prior importance for relief generalization (Weibel, 1987) and closely related to the topics discussed above.

Description by means of mathematical formulation
Intuitively, it may be assumed that the approximation of a set of points by a polynomial function, by using the least-squares method for instance, would allow the identification of line types. In fact, it turns out that, when the number of points to be approximated is large, equations become too complex for identification purposes (Maling, 1968; Buttenfield, 1985). Therefore, appproximations to define local forms will be used while limiting the polynomial degree so as to be able to use and interpret the results.

The idea retained by Affholder (1993) is thus primarily to model man-made topographic objects, such as roads, by using polynomials of the 3rd degree — cubic sections whose equations have the form $y = ax^3$ in a local coordinate system. The reason for satisfactory results in modelling roads by using such cubic curves is that the modelling method has been chosen according to the object to be modelled: the curves used are mathematically a good approximation of the clothoids used by engineers to construct roads. The point is not to find a completely generic modelling method but to find the best simple modelling adapted to a type of feature. At any rate, it can be assumed that man-made features would be easier to model than natural features for which polynomial geometry seems ill-adapted. Nevertheless, it has not been proved that, for these features, polynomial modelling would not contribute any information on characteristic shapes.

6.2.2 Spatial relations

According to Kate Beard, spatial relations between objects present constraints for each generalization operation:

> Structural constraints may be expressed in terms of maximizing, minimizing or maintaining certain relationships. In the spatial domain, distance relationships among objects are fundamental. Both interior dimensions of objects and the spacing between them are relationships which may need to be preserved or in other ways constrained. Other examples of spatial relations include direction, connectedness, containment and adjacency (Beard, 1991).

Spatial relations may be divided into relations of connexity and relations of spatial arrangement.

Connexity relations

Connexity relations should be preserved because they give information on objects which communicate in nature. Moreover, they are necessary in generalization to propagate shifting and deformation between connex objects. Besides, they allow identification of objects which share the same local geometry (e.g. a road fragment and an administrative boundary).

Topology allows expression of relations of inclusion, intersection, and adjacency between objects in a simple way. Planar graphs are often used to model topologic relations in dimension 2. Classical representation consists of memorizing, for each arc, its initial node, its final node, its right and left faces and then, for each face, the set of arcs of which it is composed. This is not the most practical representation, particularly when faces that are not directly defined are to be handled. In addition, this structure does not allow identification and immediate management of topologic objects contained in the faces (isolated nodes, dangling arcs, other faces). Topologic maps are derived from combinatorial maps (Berge, 1983), which are an improvement on the structure of classical graphs by adding a hole handling facility (David, 1991; David *et al.*, 1993). Topologic maps are based on the notion of a *dart* which is an oriented arc with functions allowing movement from one topologic object to another. These functions appear to be a convenient way of handling topologic information.

Spatial arrangement relations

Generalization needs

Proximity relations allow description of the relative location of objects one in relation to another. They contribute in several ways to generalization, the main one being rather vague as it consists of preserving the relative location of one object in relation to other ones. Studies made at the Institute for Aerospace Survey and Earth Sciences — ITC (i.e. generalization of a set of lakes, of a road network in an urban zone) made it clear that, to generalize a set of objects, it was necessary to have a great deal of information on topologic and geometric relations of objects as well as the knowledge of spatial arrangement relations obtained by computing proximity relations. The major difficulty is to preserve the structure of the object spatial distribution (Müller, 1993).

Geometric distribution of objects also allows their identification. Let us consider the generalization of a set of houses. The method consisting of increasing their size in order to make them visible and then aggregating them is unsatisfactory as it will produce erroneous information. It is necessary to preserve the geometric distribution as this will allow identification of the houses and prevent construction of a representation which looks like a set of large buildings. In terms of constraints, it is of vital importance to:

- preserve proximity; and
- preserve the geometric distribution of objects, if it is characteristic.

In terms of tools, proximity relations can be used to:

- detect a proximity conflict (or spatial conflict);
- identify a complex (structural) object;
- eliminate objects if they are part of a set of objects which are of the same nature while preserving the general distribution (structuring operator); and
- aggregate close objects of the same nature.

The problem which remains to be solved is to find out to what extent the same elements of object organization can be found at different scales. In the same way, there are critical scales at which certain objects change in dimension. Identification of spatial distribution of objects would not provide any interesting information beyond that limit. This occurs particularly when there is a significant change in scale (e.g. if a set of buildings is replaced by a single block delimited by neighbouring streets, the geometric distribution of the initial buildings does not contribute any usable information; only the density and the nature of the buildings would provide some relevant information).

Types of spatial arrangement relations

Two objects are neighbours if they are close to each other. Although simple as stated, proximity relations are difficult to use in practice. The objective is to define the way in which objects are close.

The first approach consists in qualifying the relative position of one object in relation to another. First, the positions can be described by using a particular vocabulary defined according to a frame of reference inspired by everyday language: right/left, in front of/behind, in the same place, surrounded by, parallel to. Distance can be combined with such location; it can be known precisely, defined by an interval, or unknown (Freeman, 1978; Mark and Frank, 1989; Donikian *et al.*, 1993).

The second approach is to try to define the geometric distribution of objects by means of geometric structures. Thus, objects may be:

- aligned;
- distributed orderly on a grid or network (hence, the shape of the grid or the type of network should be defined (Argilias *et al.*, 1988); it may be a TIN such as the SDS used in Chapter 8);
- distributed on a particular pre-determined structure (a circular structure with increasing density close to the centre, star-shaped structure, etc.).

Modelling proximity relations

To spot objects potentially in conflict, it is necessary to identify close objects. Many solutions can be envisaged.

- There can be a raster view of the data used, taking into account the span induced by the required symbolization, and then an identification of pixels which correspond to overlaps. The disadvantage of this method is the fact that a differentiation between real intersections and conflicts is rather complex. The problem of computation time and of data volume can also be a shortcoming,
- For each object, a computation of its neighbours is made possible through classical Euclidian methods of distance calculation, while the search for potential neighbours is optimized by means of a spatial index. In this respect, spatial indexes may be classified into classical spatial indexes (regular or recursive division of space, K-D-B tree, Grid file, R-tree, Bang file, etc. (David, 1991; van Oosterom, 1991) and topology-related indexes (Cell trees, Strip tree, Arc tree, Multi-scale line tree or BSP tree (van Oosterom, 1991); also Delaunay triangulations or Voronoï diagrams). The best choice is currently not clear. Whatever the optimization method used, the relevant information to be preserved will be the shortest distance between two objects, together with its location and direction (e.g. {dist; (x_1, y_1); (x_2, y_2)}). A qualitative representation of this information may also prove useful.

Modelling distribution structures

It is clear that there are no two identical geometrical distributions of features in nature. However, the procedure consisting in attempting to categorize distribution structures, aims at reducing their complexity so as to allow an elaboration of rules for simplification purposes of these distributions in the process of a change in scale. Consequently, our objective will be, primarily, to identify similarities in distribution.

The first step is to begin with the intuitive thought that there exist similarities in the geometric distribution of objects of a kind. Thus, searching for structures can be approached by an attempt at identifying distribution types according to objects, though it may later mean trying to define more generic structures. This procedure is feasible insofar as the structure is related to the object and, more generally, to its function. For example, a road network has an exchange function and a form composed of connected ramifications so as to cover space in the best possible way, just like irrigation networks where there is no hierarchy. Relationships between the street network and the distribution of buildings are also of particular interest since the configuration of the street network and settlements are closely related to the type of urban structure. Nevertheless, one may wonder if a unification of structures is possible insofar as the distribution structure and its generalization depend on the nature of the objects.

One of the ways of using structures may consist of defining prototypes of distribution structures, and then in attaching a set of objects to a prototype if certain membership conditions are fulfilled. It would therefore be necessary to permit certain value variations — angle, distance, number of objects — and to find a means of summarizing the result, so as to find out if a set of objects can or cannot be attached to a prototype, and if so, how. In other terms, a prototype can allow a spatial distribution type to be qualified, but its objective is not to describe each distribution precisely.

The main experimental work on prototypes was done on water networks (Argilias *et al.*, 1988) or more generally on natural structures (Wang and Müller, 1993), perhaps because the objects constructed are too diversified to allow for a geometrical structure synthesis. Nevertheless, it can be thought that certain structures are easily recognizable, such as house alignments in a zone which can be defined as a face delimited by a road network in a semi-urban area. A lot of work remains to be done in this area. This will be one of our main research themes, with a focus on complex interchanges, urban street networks and urban building structures.

6.2.3 Semantic relations

Generalization needs

The first stage in generalization consists of a transition from an initial data schema to one corresponding to a new level of perception (see Foreword above). This entails a redefinition of the relevant objects from the initial ones without modifying their geometry (or almost so, as a slight filtering operation may be performed in order to reduce the data volume). Schema generalization operators are (Ruas *et al.*, 1993): classification, aggregation (aggregation of connex objects of the same category and with the same values and attributes), generalization (putting objects into one, more generic, category) and association (computation of new relations between objects). At this stage, properties and semantic relations of objects are essentially used.

The second stage of generalization consists of modifying the geometry of objects so as to solve problems arising from symbolization. Certain conflicts cannot be resolved by a mere transformation of shapes or by displacement. Therefore, certain objects have to be aggregated, others eliminated or changes in dimension have to be made, depending on the context (e.g. San Francisco Bay in Mark (1991)). In all these cases, spatial relations and semantic relations between objects have to be taken into consideration. For example, only objects of the same nature may be aggregated; an object can be preserved because it is representative of a set of objects of the same nature or because it allows connection of objects which should be preserved. In addition, in order to move objects around, the order and the amount of displacement done will depend on the relative number of objects in conflict. So, each object (or object class) should be labelled with a displacement threshold according to the specification of the product to be made.

Types of semantic relations

The relations existing between geographic objects are relations which are defined by users according to their criteria and/or needs. Most relations are those of composition. A complex entity is composed of a set of entities which describe it. For instance, a hospital is composed of a set of buildings and areas. Such a set can be replaced by a single symbol: 'Hospital' in case of conflict or overlapping. There are other relations which often result from commmunication between two objects such as: 'gives access to', 'passes under or over'. Communication relations are important to see if an object is functional (e.g. a road cannot be eliminated if it is the only possible communication between two important objects). Other relations certainly remain to be identified. Those will, no doubt, depend on the product to be made.

Semantic relations modelling

In order to enable correct modelling of objects, the notion of simple and complex objects should be established to handle the coexistence of at least two different descriptions of the geographic world. This is, at least, a convenient if not optimal way of obtaining a double data representation in anticipation of a multiple representation (see section 6.1.4).

Semantic relations are also of essential importance to reduce the number of objects in one category. As a random elimination is not acceptable, the selection method stems either from the definition of criteria of spatial distribution or from the establishment of a hierarchy within a set of objects. From this hierarchy, each object would be given a value summarizing a set of criteria of importance according to the new specifications. In establishing a hierarchy in a set of objects, geometric structures may also be of use. Other criteria based on statistical information (Richardson and Müller, 1991) or on semantically weighted connections (Beard and Mackaness, 1993) have also been proposed; this shows the great variety of available semantic criteria.

6.3 Suggestions on modelling for generalization purposes

The aim of formalizing and modelling relations between objects is to facilitate or even make possible the generalizing procedure. This section deals with some methods used for modelling relations described in section 6.2. It is obvious that certain relations or properties are necessary for all objects (such as topologic relations) and that other relations will merely concern a restricted number of objects (e.g. house alignment). What is important is to be able to introduce properties needed for generalization into a data model while being aware of the fact that the need for modelling will be increasing, and its importance will grow together with the progress in automated generalization. The model used should therefore be easily extensible.

6.3.1 Choices in the field of modelling

By the end of 1991 the laboratory Cogit of IGN started research in order to study feasible processes for the generalization of the databases in progress at the institute. At first, it was necessary to conduct a survey on generalization (Ruas *et al.*, 1993) and then to define adequate data modelling which could be used as a basic structure in a rule-based system. Considering the progress in research on generalization and the amount of work to be done, the first step is to keep all options open: appropriate modelling which allows us to resolve the problems of generalization in the best way should be sufficiently flexible and progressive. This essential criterion has incited us to work in an object-oriented way to enable easy improvement of our modelling according to the needs and constraints encountered.

We have attempted to find a structure allowing us to carry information which we consider essential, while incorporating potentially important data even if little work has been done to confirm their pertinency (i.e. geometric structures). A set of objects of the same nature will be called a layer. The following choices were made.

1. The two fundamental groups of objects are geographic objects and topologic objects. Geographic and topologic layers are concerned.
2. A topologic layer contains the geometry of objects. In order to ensure better control of the problem of floating point approximation, the coordinates are integers.
3. Geographic objects are broken down into simple and complex objects. A simple object is related to at least one topologic object. A complex object can be composed of simple and/or complex objects.
4. All simple geographic objects corresponding to the same level of description are related to a unique topologic layer, with the exception of 'contour line objects'.[1] Thus, two geographic objects can, partially or totally, share the same geometry.
5. A topologic layer is based on a division of space into a complete planar graph. Basic topologic objects are inspired by definite objects in the theory of topologic maps. Thus, topologic objects are composed of nodes, darts and faces.
   (a) A node has either an ordered list of darts or has a surrounding face if it is pending.
   (b) A dart has its reversed dart. Each dart points to its right face and to an initial node. A dangling dart has a surrounding face: it is considered as a hole.
   (c) A face has an ordered list of darts and, possibly, a surrounding face and a set of holes.
6. A geometric layer is added to the two basic layers mentioned above. It is intended to host geometric structured primitives (such as cubic curves) which are created for generalization purposes. This layer allows description of the geometry from one or many objects without being limited by topologic relations between objects. For example, it is possible to associate a set of darts with a single geometric primitive of a cubic type or to associate a dart with a set of cubic curves and other geometric primitives to be defined. Thus, in the process of certain geometric modifications (e.g. simplification or moving around) it will be possible to make use of a geometric primitive either as a constraint or as a handy tool. Each geometric modification of a geometric primitive will lead to readjustments in the coordinates at the level of the topologic objects.
7. A structural layer is added to the other layers. This layer should allow for a description of a geometrical distribution of a set of objects. It could be used as a constraint or as a means of handling such things as the selection or displacement of objects (e.g. house alignments). This layer and the previous one are typically layers of ephemeral 'meta-information'.

---

[1]Contours are only one possible representation of relief information. They cannot be integrated, as this may result in too great a fragmentation of the basic topologic layer. However, relief should be taken into consideration for the treatment of certain data. For example, to select a water network, one could be led to establish its hierarchy by using altimetric information.

8. The four layers of objects — geographic, topologic, geometric and structural — allow description of a view on geographic data (see Foreword). Such a view will be called world or perception level. The system should contain a minimum of two worlds: that composed of the initial data, called the *world of reference*, and that which is being generalized, called the *active world*. Hence, instead of a set of representational planes associated with each individual feature (Chapter 7), we consider several worlds containing different and related images. The first approach is likely to be better suited for cartographic data storage while the authors think that the latter one is more convenient for generalization purposes.
9. The world of reference is indispensable in controlling the quality of generalization. So, the objects of the active world are linked with objects of the world of reference.
10. In order to enable a visualization of objects according to the computing means used, an extra graphic layer is necessary. Then, topologic objects are linked with graphic objects of visualization. Topologic objects have attributes which are used for deriving the graphical symbolization as well as for calculation of proximity conflicts.
11. In order to optimize proximity computation needed for conflict detection, and to handle displacement, a Quad-tree indexing technique is used. This indexing is also useful for optimization of interactive operations (e.g. mouse pointing).

6.3.2 Implementation

The implementation is based on an expert system shell (Smeci, developed by the company Ilog) which has two interesting characteristics: the inference engine allows it to generate different states and the working memory is object oriented. Thus, it is easy to define categories (object classes of Smeci), and the associated attributes allow storage of descriptive data of each category and establishment of links between objects (e.g. composition relations, topologic relations, etc.). Methods and daemons allowing handling of objects are associated with each category. These methods can be triggered by sending a message (or by an action for reflex methods). The graphic layer is managed by another software package (called Maïda2D), allowing visualization and creation of objects. These graphic objects are not saved during the different sessions. A Maïda object should be considered as a screen image of a topologic object. Each geometric computation or transformation (e.g. distance, displacement, etc.), even if triggered off interactively by means of a Maïda2D object, is computed by using the topologic objects (hence in integer coordinates), and is then sent back to the Maïda2D objects level. Furthermore, Maïda2D provides an integrated Quad-tree index which is useful for the identification of isolated objects which are contained in faces, and for the optimization of distance computation. An interface software package can also be used (Aïda); it allows release of general functions to check data, grasp parameters, etc. A diagram on modelling the active world is shown in Figure 6.1.

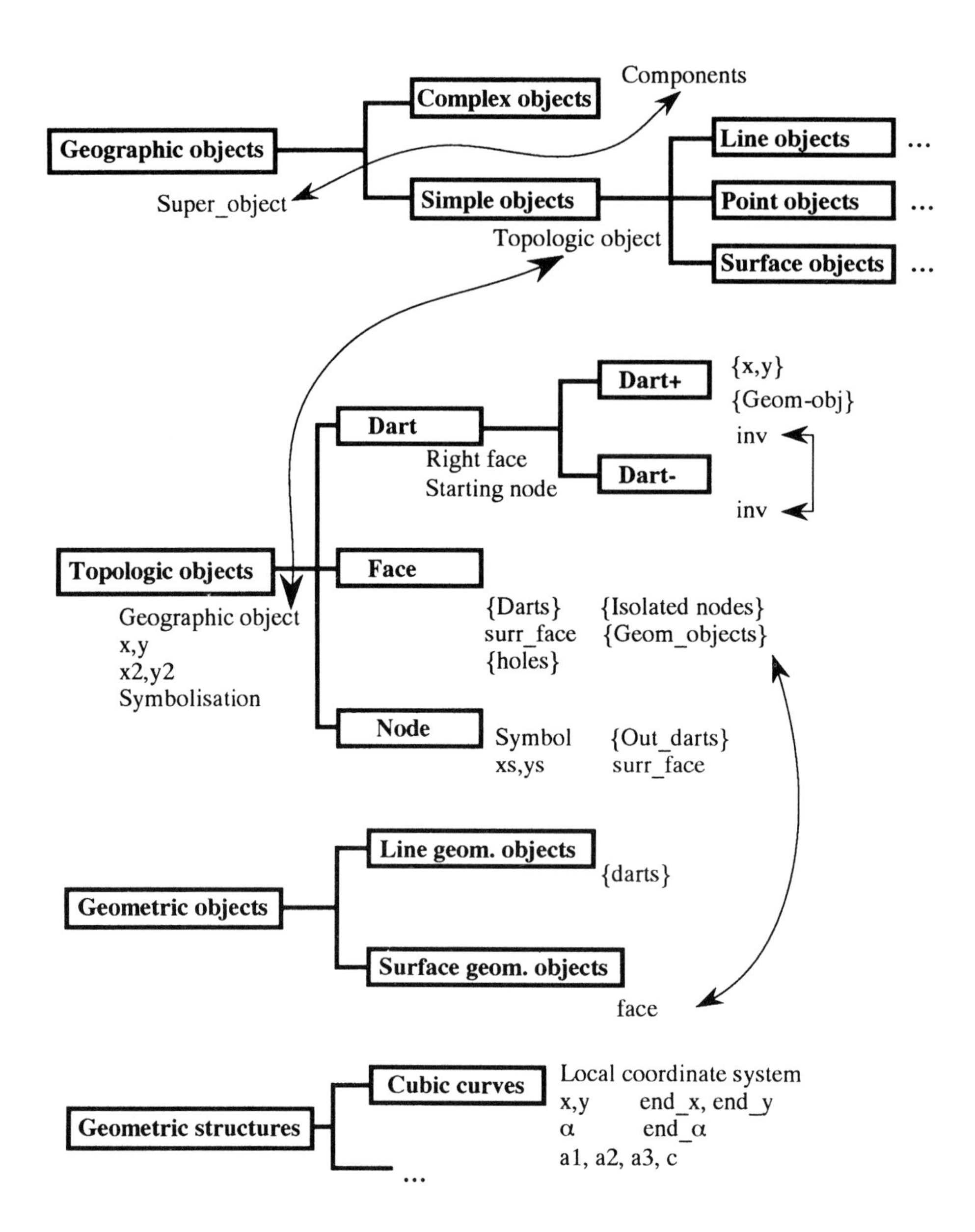

Figure 6.1 Schema of active world.

6.4 Dynamic utilization of modelling for data generalization

It is assumed that the following hypotheses and contexts hold.

- Geometric properties, semantic and spatial relations have been identified and stored in the database by means of objects (e.g. geometric objects and structures) and relations. Point and object filtering have been performed. Data is pre-generalized.
- There are two related object spaces (or worlds): the space of objects to be generalized and the space of reference.
- Objects from the active space are given a simplified symbolization allowing a visualization of overlapping between objects.

It is now time to proceed to identification and to conflict spatial resolution. What is clear is that there is no pre-defined sequence of operations. Most operations triggered off on a set of data without any control would produce violations of relations between objects which would be impossible to correct properly and rapidly. Scale-consistent symbolization, as well as new legibility constraints, will allow identification of conflictual zones. Faced with each area of conflicts, a method of spatial resolution should be identified whose objective would be to minimize the number of conflicts while respecting fundamental constraints. It is obvious that numerous methods of spatial resolution could be envisaged for each conflict. Rules should enable the selection of the best possible methods. The mechanism of state creation and backtracking allow the user to go back to one or many previous states in the case of a deadlock or of an increase in the number of conflicts. Insofar as the objective is to handle only one database, operations will be essentially sequential.

It is already possible to assume that displacement should be controlled from the reference objects and by constraints of global forms.

Studies of the progress at IGN aim at refining data modelling and at analysing optimal utilization of generalization operators as well as their demand in the field of modelling. Table 6.1 shows correlations between classical operators of generalization (i.e. selection, shape simplification, caricature, aggregation, etc.) and data modelling.

6.5 Conclusions

A lot of work has been carried out in the following areas:

- geometric operations and, to a lesser extent, geometry characterization,
- cartographic/generalizing knowledge, but most often limited to 'surface rules' or to narrow domain knowledge bases,
- models of generalization, but which are still far from being implemented.

The authors think that in order to cope with the holistic nature of the generalization process, especially for medium scales, and to develop more general rules, a lot of attention has to be paid to information/knowledge entailed in initial data and which influences generalization decisions.

In this paper the authors have tried to summarize the different information/knowledge categories that can be considered and some possible directions for their representation and use in a generalization process. It is clear that the final word has not been said with

Table 6.1 Generalization operators and object properties and relations

Property operation	Geometry	Connexity	Proximity/inclusions structures	Semantic properties
Selection	T: Elimination of too small objects	T: Selection of objects which implement a connexity link		T: Importance of object depends on their nature
Filtering	C: Maintenance of characteristic local shapes	C: Filtering applied between topologic nodes		
Smoothing	C: Maintenance of characteristic local shapes T: The geometric class dictates the choice	C: No creation of new intersections	C: Relative locations are to be maintained	
Caricature	C: Maintenance of characterstic local shapes T: The geometric class dictates the choice of the algorithm	C: No creation of new intersections	C: Relative locations are to be maintained	
Aggregation	C: Maintenance of characteristic local shapes The geometric class of the resulting object has to be the same as before	C: Topologic changes must be restricted to aggregated objects T: Adjacent objects may be aggregated	C: The new objects must be contained in the same face T: Close objects may be aggregated	C: Objects of similar nature may be aggregated T: Components of a complex object can be aggregated
Structuration or typification		C: Topology must be maintained	T: The structure is represented and generalized	C: Objects of close nature may belong to the same structure
Cusp/collapse	T: Detection of small objects	C: Topology update	C: Update of structure of proximity relations	T: Definition of applicable symbolization
Enlargement	C: Maintenance of characteristic local shapes T: Detection of small objects	C: No new intersections	C: The enlarged object must remain in the same face	
Displacement	C: Maintenance of characteristic global shapes T: Irregular shapes have to be moved first	C: Angles of intersections must be maintained T: Propagation throughout the network	C: Maintenance of distribution structures T: Propagation on close objects	C: Important objects are displaced by others

T: Relation/property is used as a tool.
C: Relation/property acts as a constraint.

respect to these issues but, still, our claim is that it is possible to implement some of these ideas provided that the development environment is flexible and extensible enough to allow for further additions and refinements. The system under development, based on an OO and KB shell, already incorporates a significant part of the ideas expressed in this paper.

References

Affholder, J.G., 1993, Road modelling for generalization, in *Proceedings of the NCGIA Initiative 8 Specialist Meeting on Formalizing Cartographic Knowledge*, Buffalo, pp. 23–36.

Argilias, D., Lyon, J. and Mintzer, O., 1988, Quantitative description and classification of drainage patterns, *Photogrammetric Engineering and Remote Sensing*, **54**(4), 505–9.

Ballard, D., 1981, Strip-trees: a hierarchical representation for curves, *Comm. of the ACM*, **14**, 310–21.

Beard, K., 1991, Constraints on rule formation, in Buttenfield, B.P. and McMaster, R. (Eds) *Map Generalization: Making Rules for Knowledge Representation*, pp. 121–35, London: Longman.

Beard, K. and Mackaness, W., 1993, Graph theory and network generalization in map design, in *Proceedings of ICC'93*, Cologne, pp. 352–62.

Berge, C., 1983, *Graphes*, Paris: Gauthier-Villars.

Bjorke, J. and Midtboe, T., 1993, Generalization of digital surface models, in *Proceedings of ICC'93*, Cologne, pp. 363–71.

Brassel, K.E. and Weibel, R., 1988, A review and conceptual framework of automated map generalization, *IJGIS*, **2**(3), 229–44.

Buttenfield, B.P., 1985, Treatment of the cartographic line, *Cartographica*, **22**(2), 1–26.

Buttenfield, B.P., 1991, A rule for describing line feature geometry, in Buttenfield, B.P. and McMaster, R.B. (Eds) *Map Generalization: Making Rules for Knowledge Representation*, pp. 150–71, London: Longman.

Buttenfield, B.P. and McMaster, R.B. (Eds), 1991, *Map Generalization: Making Rules for Knowledge Representation*, London: Longman.

Cheylan, J.P., 1989, Chiffres et cartes: une réunion réfléchie, *GIP Reclus Technical Report*, Montpellier.

David, B., 1991, 'Modélisation, représentation et gestion d'information géographique', unpublished PhD thesis, University Paris VI.

David, B., Raynal, L., Schorter, G. and Mansart, V., 1993, GeO2: Why objects in a geographical DBMS? in *Proceedings of SSD'93*, Singapore, June 1993, in Abel, B. and Doi, B.C. (Eds) *Advances in Spatial Databases* LNCS 692, pp. 264–76, Springer Verlag.

Donikian, S., Hégron, G. and Arnaldi, B., 1993, A declarative design method for 3D scene sketch modeling, in *Proceedings of Eurographics*.

Douglas, D. and Peucker, D., 1973, Algorithms for the reduction of the number of points required to represent a digitised line or its caricature, *The Canadian Cartographer*, **10**(2), 112–23.

Dutton, G., 1981, Fractal enhancement of cartographic line detail, *The American Cartographer*, **8**(1), 23–40.

Freeman, H., 1978, Shape description via the use of critical points, *Pattern Recognition*, **10**, 159–66.

Gouyet, J.F., 1992, *Physique et Structures Fractales*, Paris: Masson.

Hoffman, D.D. and Richards, W.A., 1982, Representing smooth plane curves for visual recognition, in *Proceedings of AAAI'82*, pp. 5–8.

Lanza, L. and La Barbera, P., 1993, Hydrologically oriented geographical systems: scale problems and reliability of data, in *Proceedings of EGIS'93*, Genoa, pp. 653–62.

Leyton, M., 1988, A process grammar for shape, *Artificial Intelligence*, **34**, 213–47.

Lowe, D.G., 1988, Organization of smooth image curves at multiple scales, in *Proceedings of the 2nd ICCV*, pp. 558–67.

Maling, D.H., 1968, How long is a piece of string? *The Cartographic Journal*, **5**(2), 147–56.

Mark, D., 1991, Object modelling and phenomenon-based generalization, in Buttenfield, B.P. and McMaster, R.B. (Eds) *Map Generalization: Making Rules for Knowledge Representation*, pp. 103–18, London: Longman.

Mark, D., 1992, 'The importance of a cognitive science perspective for the design of geographic databases', presented at ESF GISDATA Workshop, Aix-en-Provence.

Mark, D. and Frank, A., 1989, Concepts of space and spatial language, in *Proceedings of Auto Carto 9*, Baltimore, pp. 538–55.

McMaster, R., 1987, The geometric properties of numerical generalization, *Geographical Analysis*, **19**(4), 330–46.

Milios, E.E., 1989, Shape matching using curvature processes, *Computer Vision, Graphics, and Processing*, **47**, 203–26.

Mokhtarian, F. and Mackworth, A.K., 1992, A theory of multiscale, curvature-based shape representation for planar curves, *IEEE Transactions on Pattern Analysis and Machine Intelligence*, **14**(8), 789–805.

Müller, J.C., 1986, Fractal dimension and inconsistencies in cartographic line representations, *The Cartographic Journal*, **23**, 123–30.

Müller, J.C., 1993, Procedural, logical and neural net tools for map generalisation, in *Proceedings of ICC'93*, Cologne, pp. 181–91.

Piron, M., 1993, Changer d'échelle: une méthode pour l'analyse des systèmes d'échelles, *L'espace géographique*, **2**, 147–65.

Plazanet, C., Affholder, J.G., Lagrange, J.P. and Ruas, A., 1994, Représentation et analyse de formes linéaires pour l'automatisation de la généralisation cartographique, in *Proceedings of EGIS'94*, Vol. 2, Paris, pp. 1112–21.

Richardson, D. and Müller, J.C., 1991, Rule selection for small-scale map generalization, in Buttenfield, B.P. and McMaster, R.B. (Eds) *Map Generalization: Making Rules for Knowledge Representation*, pp. 136–49, London: Longman.

Rosin, P., 1993a, Multiscale representation and matching of curves, *CVGIP: Graphical Models and Image Processing*, **55**(4), 286–310.

Rosin, P., 1993b, Non-parametric multi-scale curve smoothing, in *Proceedings of SPIE Conference, Application of AI, XI: Machine Vision and Robotics*, pp. 66–77.

Rosin, P. and West, G.A.W., 1990, Segmenting curves into elliptic arcs and straight lines, in *Proceedings of the 3rd ICCV*, pp. 75–8.

Ruas, A., Lagrange, J.P. and Bender, L., 1993, Survey on generalization — Etat de l'art en généralisation, *IGN Technical Report*, DT 93-0538.

Thapa, K., 1989, Data compression and critical points detection, in *Proceedings of Auto Carto 9*, Baltimore, pp. 78–89.

van Oosterom, P., 1991, Reactive data structures for geographic information systems, unpublished PhD thesis, University of Leyden.

Wang, Z. and Müller, J.C., 1993, Complex coast line generalization, *Cartography and Geographic Information Systems*, **20**(2), 96–106.

Weibel, R., 1987, An adaptative methodology for automated relief generalisation, in *Proceedings of Auto Carto 8*, pp. 42–9.

Weibel R., 1991, Amplified intelligence and rule-based systems, in Buttenfield, B.P. and McMaster, R.B. (Eds) *Map Generalization: Making Rules for Knowledge Representation*, pp. 172–86, London: Longman.

Williams, D. and Shah, M., 1992, A fast algorithm for active contours and curvature estimation, *Image Understanding*, **55**(1), 14–26.

7

Object-oriented map generalization: modeling and cartographic considerations

Barbara P. Buttenfield

NCGIA, State University of New York, Buffalo, New York 14261, USA

National Mapping Division, United States Geological Survey, 519 National Center, Reston, Virginia 22092, USA

7.1 The problem

> There has been much excitement about the introduction of object-oriented programming in GIS . . . The object-oriented environment, where procedures (methods) are bound to the object, objects communicate to each other and inherit attributes and methods from others, seems to offer great potentials for implementing generalization procedures. The concept of "delegation between objects", in particular, could be used to perform updates concurrently across all map-scale layers in the database. . . . the proposed models are attractive but have no proven records yet in the field of generalization (Müller *et al.*, 1993, p. 5).

In generalization, the logical association between cartographic features and their geographic counterparts must be preserved at all map scales, regardless of scale-dependent characteristics that epitomize the nature of most geographic phenomena. Scale-dependent characteristics may be metric, as in measures of distance or direction, which have been demonstrated to change with resolution (Richardson, 1961). Non-metric characteristics include topological relationships or other attributes (e.g. place names may have meaning at some but not all scales). Preserving the association between geographic and cartographic meaning is the responsibility of the cartographer who compiles the feature. It is also the responsibility of the cartographer who modifies the feature during scale change or other generalization operations. Historically, these responsibilities have been carried out using intuition, cartographic expertise, and the cartographer's knowledge of geographic process and form. One finds evidence that digital representation should be carried out mimicking the manual process. For example, Richardson and Müller (1991, p. 138) comment that automated solutions to generalization should emulate the cartographer's decision making. However, it is not clear that algorithms and procedures can by themselves suffice.

This paper argues that digital cartographic representation requires robust algorithms *and* a suitable data model. 'The way the data model is organized and can be generalized is likely to influence the performance of cartographic generalization'. (Müller *et al.*, 1993, p. 6). Agreed, there are requirements for encoding symbols on the map, and these may be formulated in the form of production system rules, frames, or parameterized expressions (Adeli, 1990) to select feature color, shading pattern, typeface, and so on, responsibly. But before the marks are made on the page or the CRT screen, information must be resident in the database to attach location and meaning to the marks, and to keep in mind '. . . the intimate relationship between generalization at the modeling level and generalization at the "surface" (e.g. graphical representation)' (Müller *et al.*, 1993, p. 6). Nyerges (1980, 1991) has previously referred to the non-graphic digital data model that contains spatial and attribute relationships as 'deep structure', and distinguished this from the graphic image, which he calls 'surface structure'. In this paper, operations that manipulate the data model will be referred to as 'deep' and the resulting organization and form as 'deep structure'. Operations that affect the cartographic depiction will be referred to as 'surface' and the organization and form of the cartographic display will be referred to as 'surface structure'.

In addition to supporting 'surface' generalization, the data model must support operations that precede map display, that is, generalization at the deep or modeling level. Deep or modeling operations will include formation of new data or data layers, for example spatial buffering (a form of feature-based classification), feature intersection (relevant to placement and displacement), or may involve results of analysis and measurement, e.g. distance computation for tolerance value selection and modification. To the extent that data representations vary with scale, algorithms underlying these operations may vary. It will be shown in this paper that data collection, compilation and/or delineation also have an effect on surface meaning.

The current approach to dealing with multiple representations at many National Mapping Agencies (NMAs) is to generate multiple databases, and utilize a particular database to print maps at a particular scale. This effectively limits a single deep meaning to a single surface meaning, in the following manner. Representations of the same feature at different scales are not linked, since the databases are not linked. This multiple database approach works when the agency mission prioritizes creation and revision of paper map series. Problems arise when the mission scope expands, as is happening in many countries. USGS National Mapping Division's mission has been revised recently to provide not only paper maps but digital geographic and cartographic data, and to conduct research responsive to national needs. Accomplishing this mission requires maintenance of '. . . a digital geographic/cartographic data base for multi-purpose needs [to assist] Federal and State agencies in developing and applying spatial data' (USGS, 1993, section 6.1).

The database must continue to support production and printing of paper map series, as well as supporting spatial queries for GIS mapping applications. However, it is not clear that a data model designed to print paper map series one scale at a time can support a mission of multi-purpose research needs. First, GIS analytical results may vary with scale, as, for example, in the case of the Modifiable Areal Unit Problem (Openshaw, 1984). Other disadvantages of the multiple database approach relate to costs of redundant data production, costs of redundant data revisions, and costs of preserving consistency across multiple databases. On the other hand, it does seem reasonable that a database designed for multiple scales of analysis should also be able to support map series production at isolated scales.

> Solutions to these impediments require moving beyond the paradigm of traditional paper map series, without sacrificing support for the production of paper maps. That is to say, the generation of digital products can no longer be driven by paper map production, as the needs for spatial data have become much broader and complex. Generalization facilities must be provided by GISs to support the use of geographic information at multiple scales for multiple purposes and tasks (Müller *et al.*, 1993, p. 3).

This paper proposes an object-oriented multi-scale data model based upon the Digital Line Graph (DLG-E) data model developed at the US Geological Survey. This paper extends the DLG-E model to incorporate multiple representation schemes for a single object record within a single database, thus accomplishing a link from a single deep meaning to multiple surface meanings. The paper covers encapsulation and polymorphism for the extension, with emphasis on representational rules relevant to scale-changing operations. Encapsulation in object-oriented programming refers to combining procedures with a data record, literally embedding rules for object behavior into the object description. Polymorphism refers to embedding the procedures in such a way that parts and subparts of a complex object share a procedure but apply it with differing results. For scale changing, this could take the form of simplifying some but not all branches of a braided stream, for example, or deleting buildings but not transportation routes for an industrial complex. The case studies given in section 7.4 will demonstrate one way to implement these capabilities, following a discussion of data models in general, and the DLG-E data model specifically.

7.2 *The DLG-E data model*

Peuquet (1984) describes a data model as a level of data abstraction, which is itself a form of generalization (McMaster and Shea, 1992). A data model specifies the components and relationships among components abstracted from the real world. Also included are a set of integrity constraints, or 'axioms' concerning the components and relationships. A data model is not necessarily independent of specific data structures that may be generated from it (Guptill, 1990). Data structures specify the logical organization of data model components and how the relationships are to be explicitly defined. (Both of these are distinct from file formats that specify the implementation of a data structure in a particular computing system environment.)

Methods for data modeling are often discussed within the larger context of knowledge representation, and the concepts have a mathematical basis in set theory (Lagrange *et al.*, 1993). The DLG-E data modeling framework is drawn from Borgida (1986), and includes definitions similar to those found in the *Spatial Data Transfer Standard* (FIPS, 1992). In that framework, a data model contains objects, attributes, and relationships (these are the components, or basic units for representing knowledge). These components correspond to entities in the real world (adapted from Guptill, 1990).

- An **entity** is a real-world phenomenon that cannot be subdivided into phenomena of the same kind.
- An **object** is a digital representation of the entity, and has associated locational and non-locational **attributes** or characteristics.

- **Relationships** are topological and non-topological links between objects having similar attributes. Groups of objects can be linked to form a **class**. A class is also an object.
- A **feature** is a class of objects with common attributes and relationships. The concept of a feature encompasses both entity and object.

As Guptill (1990, p. 18) states, 'Features are the sum of our interpretations of phenomena on or near the Earth's surface.' The DLG-E data model is feature based, and in this context the term feature refers both to the real phenomenon (the entity) and to the representations of knowledge about it (the objects, attributes, and relationships). As these representations vary with scale, there may be many digital versions corresponding to a single entity. Guptill (1990, p. 2) states 'The underlying philosophy of DLG-E is to view cartography as an information transfer process that is centered about a spatial data base which can be considered, in itself, a multifaceted model of geographic reality. DLG-E data form the contents of the spatial data base. The DLG-E features must be representative of a model of geographic reality.'

To be representative of a model of geographic reality that varies with scale, one must anticipate the variations in objects, attributes or relationships that may be encountered in the digital representation. Changes in object geometry have been documented (Richardson, 1988, 1989; Müller, 1990; Buttenfield *et al.*, 1991; João *et al.*, 1993; Leitner, 1993). Attribute names vary as well. For example, a 'divided highway' may be categorized at a finer resolution as a 'toll road' or as a 'freeway'. At still another resolution, it might simply be categorized as a 'limited access road'. Third, one must attend to variations in topological and non-topological relationships, to protect against possible intransitivities that may occur during scale change (Corbett, 1979; Bruegger and Frank, 1989). To accommodate these variations, a multi-scale data model should incorporate multiple object representations linked to individual entities.

Returning for a moment to surface meaning and deep meaning, the data model and its relationships can be thought of as providing a mechanism to link geographical meaning ('this is a braided stream and that over there is an industrial plant') with deep meaning (the digital data definitions and the formalized relations, such as stream order, or components of the industrial plant). Links between deep meaning and surface meaning are established through rules of data modeling (compilation, delineation, and GIS operations) and data depiction, what have been respectively referred to by Brassel and Weibel (1988) as statistical and cartographic generalization. Geographical meaning does not change; that is, a braided stream remains a braided stream. Deep meaning may change through data analysis or manipulation (as in creating a new data layer by polygon intersection, or modifying an attribute set by reclassification). Surface meaning will change most often, through changes to symbolization. Changes to both deep and surface meaning fall under the rubric of generalization.

The remainder of this paper outlines extensions to the DLG-E data model that support scale-induced changes in object geometry, changes in attributes, and changes in feature relationships. Object-oriented methods for data modeling will inform the discussion, which will be clarified using two examples. The first is a cultural feature, an industrial complex of buildings comprised of different components. The second is a natural feature, a stream changing from a braided channel at larger scales to a single channel. Both types of features occur commonly on National Mapping Division Maps, and vary in geometry and topology with the scale of their map representation. A Digital Orthophoto Quadrangle is available (compilation at 1:24 000) for control of feature descriptions in the extension. Obviously, two case studies taken from only two domains (hydrography and built-up land) cannot by themselves justify extending the DLG-E data model. The intention here is to initiate discussion, and to determine whether this research thread is justifiable for further pursuit.

7.3 Scale-dependent extensions to the DLG-E data model

The DLG-E data model is both hierarchical and object oriented. Guptill (1990) refers to it as 'feature-oriented', to acknowledge that both real world entities and digital objects are encompassed within the data model. Hierarchical relationships lend themselves well to object orientation as they facilitate encapsulation and inheritance (Parsaye *et al.*, 1989). The DLG-E hierarchy begins with a partition of five Views on the database, including Cover (physically evident areal extent), Division (administrative or political boundary), Ecosystem (defined by natural covers, e.g. wetlands), Geoposition (geodetic control), and Morphology (defined by landscape physiography). Both examples in this paper are taken from the view called Cover.

A DLG-E feature specification takes a general form (Guptill, 1990). The multi-scale extension can follow the same form. Each object definition includes a list of attributes and attribute values. Locational attributes include the spatial and dimensional components; and non-locational attributes include categories refining the object definition. Relational links between objects are defined by the dimensionality of the objects being related, e.g. 'is-composed-of' relates a feature to features of all dimensions, 'is-within' relates a point or chain to a polygon, and so forth (Guptill, 1990). These relational links allow feature definitions for compound objects, as, for example, in the case of an industrial complex.

The DLG-E data model defines features and attributes and associates rules for feature delineation, data extraction and representation, according to production specifications at the USGS. (Representational rules will be emphasized in this paper.) Representational rules include pen width, color, single- or double-line symbol specification, or symbol dimensionality (point, node, chain, area), etc. These are surface rules. The DLG-E data model is currently designed to incorporate only a single scale product within a database, and a single set of surface rules. Examples will be presented in the case studies in section 7.4 which incorporate multiple sets of surface rules.

It is logical to ask what relationship between surface meaning and deep meaning is served by scale-specific data models. The implication of any map symbol is that its representational rules provide for a graphical depiction that is logical and realistic with respect to the geographical meaning and by inference to deep meaning as well, since the data model has an impact upon what surface representations may be generated from it. A grammatical example of surface and deep meaning can be provided by the two sentences 'The boy threw the ball' and 'The ball was thrown by the boy'. Though syntax differs, the deep meaning is the same. Likewise, though graphical depiction of an industrial complex may vary with map scale, the intention is that map readers will identify the geographic entity nonetheless as an industrial complex. The implication of a scale-specific data model is that deep meaning is also scale specific (at very small scales, some buildings may not be resolved, or some railway sidings may not be visible, etc.), even though the geographic entity meaning remains the same (the industrial plant is an industrial plant at all levels at which it or its components can be resolved).

The extension to the data model proposed here incorporates multiple sets of representational rules, based upon standard procedures and specifications for symbol appearance on topographic maps at standard scales (at USGS, these are 1:24 000; 1:62 500; 1:100 000; and 1:250 000). Despite the fact that digital data are produced at only two of these scales, inclusion of all four scales is intended for proof of concept. It has been shown that map features undergo only slight change at some scales, and catastrophic change at others (Müller, 1990; Buttenfield, 1991; Buttenfield *et al.*, 1991; João *et al.*, 1993). Conventional wisdom indicates that these changes are introduced by the cartographer to reflect changes in details apparent at some but not all map scales. In a National Mapping Agency context, catastrophic changes are codified in the compilation of data at isolated scales. It is difficult to determine whether the base scales of 24 000, 100 000, and so on, are selected for pragmatic reasons or due to assumptions that these are the catastrophic scales.

Differing geographic processes operate at differing scales. For example, isostatic rebound following glacial retreat is evident at very small scales (e.g. 1:1 000 000 or smaller) while erosion and deposition are evident at larger scales (e.g. 1:20 000 or larger). Because of the scale-dependent nature, it seems reasonable to expect that different types of feature domains (transportation, or hydrography) should undergo representational change at different scales. For efficiency in the data model, additional sets of representational rules are not needed except at catastrophic scales, which have not been identified for specific feature domains. Nor has a paradigm been established by which such identification can proceed. With the exception of early work by Imhoff on mapping settlements, this line of research has not been pursued. For the time being, then, one can only infer where catastrophic changes occur, by studying map features represented at all existing map scales. In that context, the inclusion of all four scales in the case studies in section 7.4 is intended to reflect existing cartographic evidence of representational rules for USGS topographic map series. It serves the purpose of data modeling, and proof-of-concept for such a paradigm, as opposed to demonstrating that these four scales will be critical for identification of catastrophic change. Surface meaning (representational rules) will be emphasized in the case studies, and the point of deep meaning will be returned to in section 7.5.

7.4 DLG-E case studies

7.4.1 DLG-E case study 1: industrial complex

The existing DLG-E data model components applicable to an industrial complex (an aluminum plant located near Vancouver, WA, USA) are defined below. The aluminum plant is an entity. It cannot be subdivided into other phenomena of the same kind (that is, into other aluminum plants). Its digital representation is composed of objects, attributes, and relationships, grouping buildings and structures serving various purposes, including manufacturing, storage tanks, a smoke stack, warehouses, material transport, and a cooling pond. The roads and railway sidings might be considered part of the plant or as a portion of the railway web, depending on how the data is modeled. This point will be returned to in the discussion below.

Figure 7.1 illustrates the locational components of the aluminum plant at a number of USGS standard scales, as well as a Digital Orthophotoquad (DOQ) depiction of the same industrial complex. Publication dates for all maps and for the DOQ range from 1974–9, with the exception of the 1:62 500 version (last revised in 1963). In comparison

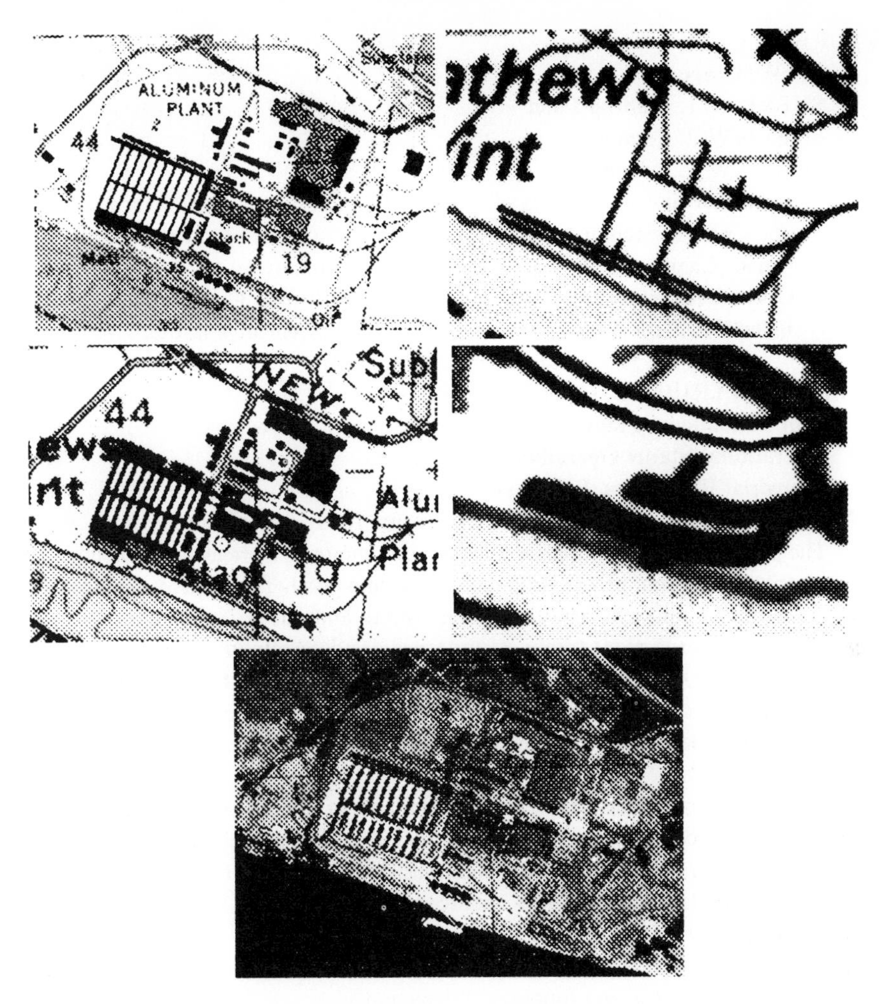

Figure 7.1 An aluminum plant near Vancouver, WA. Upper left, 1:24 000; middle left, 1:62 500; upper right, 1:100 000; middle right, 1:250 000, bottom is taken from a Digital Orthophotoquad. All images were scanned from the original maps at 150 dpi (vertical and horizontal) and 24-bit color, then converted to TIFF format and enhanced for clarity using Adobe Photoshop. All images are shown at 1:24 000 scale.

with the 1:24 000 map, one can see little difference in the plant during this time period. One can also see that the aluminum plant is represented by different object sets at different scales.

The DLG-E data model might represent the relevant digital knowledge as follows (definitions and categories taken from Guptill, 1990). The format given here is standard for all DLG features, in terms of View, Subviews, Attributes and Attribute Value Lists, etc. Definitions (such as industrial site) are taken directly from the DLG Circular (Guptill, 1990). Extensions to the DLG-E data model include merging rules for multiple scales into a single data model. Symbol type descriptions are taken from USGS topographic map legends, and font and font size are taken from topographic maps.

FEATURE

VIEW: **Cover**

SUBVIEW 1: **Built-up Land**

SUBVIEW 2: **Complex**

SUBVIEW 3: **Industrial**

Industrial Site = A group of associated structures functioning as a unit used for refining a material or manufacturing a product

Name (Character identifier, Not applicable, Unknown)

Operational Status (Abandoned, Operational, Under Construction, Unknown)

Industrial Plant Type (Ammunition, Cannery, Cement, Filtration, Refinery, Sawmill, Sewage Disposal, Unspecified)

ATTRIBUTE/ATTRIBUTE VALUE LIST

Name **Aluminum Plant**

Operational Status **Operational**

Industrial Plant Type **Refinery**

DELINEATION (ground truth — what the feature looks like on the ground)

The plant may be composed of buildings, warehouses, storage and associated structures lying adjacent to a cooling tank or cooling pond. There may be transportation paths (roads, railways, or water channels) into and out of the plant area.

The limits of the plant (plant boundaries) are determined by the outer extent of land covered by buildings associated with plant activity. If a fence line is present, this may be used to delineate boundaries. If no fence line is present, then the limits of broken or open ground may be used to infer a boundary.

DATA EXTRACTION

Data extraction specifications are not available at the time of this writing, and will not be covered, but should follow USGS production specifications for 1:24 000 compilation including Capture Conditions, Source Interpretation guidelines (image sources, graphic sources, DLG-3 sources, and others), and Valid Attributes.

REPRESENTATION RULES

Composition rules (Relationships between structures must be defined here, and related to geometric and topological depiction.)

If industrial site is composed of more than one structure, then feature is to be composed according to the rules outlined below. Individual structures are to be represented according to the composition rules (at scale) for those structures.

Vertical and horizontal topology should be preserved.

If industrial site is composed of roads and rail lines then roads are composed of ordered chains, vertically related to other features.

Product Generation rules (definition of feature instance based on global rules)

1:24 000

The attribute name will be shown in 6 point type, sans serif, all caps. If the industrial site is composed of buildings and other structures, then

buildings will be shown as black-filled standard square symbols.

warehouses shall be shown as open (not filled) symbols.

storage tanks will be shown as filled circle symbols.

smoke stacks will be shown as an open circle with a central dot.

The attribute name ('Stack') will be shown in 5 point type, sans serif, caps/lower case.

substations will be shown in proper location and orientation by a dashed line. The attribute name ('Substation') will be shown in 6 point type, sans serif, caps/lower case.

cooling ponds will be shown in proper orientation.

If the industrial site is composed of roads and rail lines, then delimit instances using rules for network links. Also,

all roads will be shown.

all rail lines will be shown.

1:62 500

The attribute name will be shown in 6 point type, sans serif, caps/lower case.

Other delimiting rules as for 1:24 000.

1:100 000

If the industrial site is composed of roads and rail lines, then delimit instances using rules for network links. Also,

all roads will be shown.

all rail lines will be shown.

1:250 000

If the industrial site is composed of roads and rail lines, then delimit instances using rules for network links. Also,

all rail lines will be shown.

Special conditions at all scales (specific definition rules for this feature instance)

buildings > 0.02 inches shall be shown with explicit shape and orientation.

buildings < 0.02 inches shall be shown as standard square symbols.

roads and rail lines < 0.02 inches in length will not be shown.

Relational tables are associated with the data model components to link spatial and non-spatial attribute values for individual structures in the industrial site in the current version of the DLG-E data model, and need not be newly constructed for the extension. One of these (the Feature Object Table) summarizes attribute definitions, relationship definitions, attribute values, and relationship instances for each object, and includes an identifier (a feature object key) to facilitate access to coordinates, etc. Seven other relational tables specify attribute definitions, attribute (major and minor) codes, and define topological and non-topological relationships between objects.

7.4.2 DLG-E case study 2: braided stream

The existing DLG-E data model components and attribute values for a stream (Salmon Creek, a braided stream near Vancouver, WA, USA) are defined below (Guptill, 1990). This differs from the previous case study in several respects. First, the digital representation is not composed of various structures, but a single feature whose geometry is complex. With scale change, the braided channel (which is encoded as an attribute) will first be represented explicitly and then implicitly, as a single channel.

Figure 7.2 illustrates the locational components of the braided stream at USGS standard scales. A hand-compiled version at all scales is published in Buttenfield (1989). The source map sheets are identical to the sources for the aluminum plant case study, except that a stream representation is also included on the 1:500 000 map series. This figure demonstrates that scale-dependent variations exist, and that the catastrophic changes in geometry occur between 1:100 000 and 1:250 000.

The relevant digital knowledge is represented in the DLG-E data model as follows. Once again, the format given here is standard for all DLG features, in terms of View,

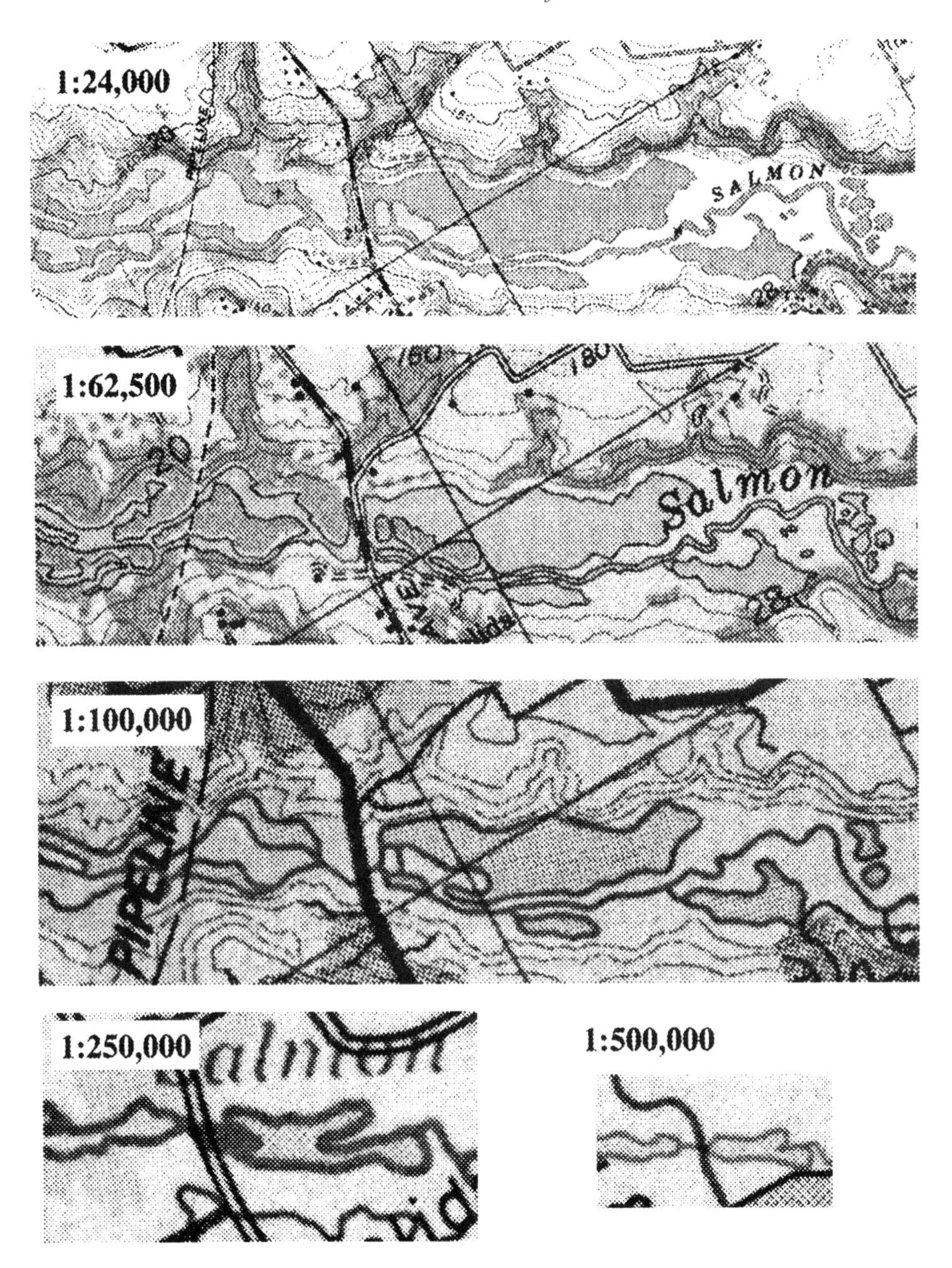

Figure 7.2 *A braided stream (Salmon Creek) north of Vancouver, WA. Images were scanned from sources at 1:24 000; 1:62 500; 1:100 000; 1:250 000; and 1:500 000 but all are shown at 1:24 000 except for the smallest scale images, which could only be enlarged to 1:50 000 and 1:100 000, respectively.*

Subviews, Attributes and Attribute Value Lists, etc. Definitions and attribute categories are taken directly from the DLG Circular (Guptill, 1990), as are the numeric threshold values given for compilation, extraction, and representation. Scale-dependent extensions of representational rules are original to this paper.

FEATURE

VIEW: **Cover**

SUBVIEW 1: **Water**

Stream/River = A body of flowing water.

Delineation Status (Apparent, Definite, Unsurveyed)

Elevation (Vertical Datum Category, Stage, Unspecified, Unknown)

Directional Status (Alternating, Bidirectional, Not Applicable, One Way, Unknown, Unspecified)

Hydrographic Category (Intermittent, Perennial, Unknown)

Hydrographic Form (Braided, Not braided)

Name (Character Identifier, Not applicable, Unknown)

ATTRIBUTE/ATTRIBUTE VALUE LIST

Delineation Status **Definite**

Elevation **Vertical Datum Category** = Mean Sea Level

Stage = Normal Pool Elevation

Directional Status **One Way**

Hydrographic Category **Perennial**

Hydrographic Form **Braided**

Name **Salmon River**

DELINEATION (ground truth — what the feature looks like on the ground)

The upper limit of a stream is determined to be the point at which the feature first becomes evident as a channel.

The limits of a perennial stream are determined to be the height on the banks when the water is at average water elevation.

The limits of an intermittent stream are determined to be the height of the banks when the water is at high water.

The lower limit of an areal stream entering into bays, estuaries, gulfs, oceans, or seas is defined at the place where the stream reaches a width of one nautical mile with no further constrictions, if the land and water do not otherwise make the division obvious.

DATA EXTRACTION (this is based upon extraction at 1:24 000 scale)

Capture Conditions

All streams are captured to within 0.5 inches of saddles or divides.

All streams intersecting sheet boundaries will be captured.

All streams connected to springs or ponds will be captured.

A representative pattern will be captured for braided streams: perennial channels will always be captured, as will double-line streams and streams which form bounding limits.

All tributaries are captured having a length of 1.25 inches or longer.

Tributaries shorter than 1.25 inches are captured if needed for contour portrayal.

Source Interpretation Guidelines

(All sources)

If a stream is controlled by locks, a normal pool elevation is required. Water surface elevations are given to the nearest foot. Artificial lakes formed by a dam are considered under the feature LAKE/POND. Dry washes, gulches, and ephemeral streams are considered under the feature WASH.

(Image sources)

If an image shows a lower than average water elevation, the mean water level must be obtained from ancillary sources.

(Graphical sources)
Approximate, indefinite, and unsurveyed streams/rivers have a delineation status of 'apparent'. For a braided stream, all channels will be captured.
(DLG-3 sources)
Approximate, indefinite, and unsurveyed streams/rivers have a delineation status of 'apparent'.
Valid Attributes
All attributes are valid.

REPRESENTATION RULES

Composition rules (Relationships between structures must be defined here, and related to geometric and topological depiction.)
If a stream is linear, stream is composed of sequenced, directed chains.
If a stream is areal, stream is composed of area bounded by shoreline(s), with inflow/outflow from junction, vertically related to other features.
Product Generation rules (delimit instances using rules for network links.)

1:24 000

The attribute name will be shown in 6 point type, sans serif, italic and blue color, following the general trend of the primary channel.
If a braided channel, then the stream channel will be shown in a blue double-line symbol with a 30% cyan screen tint (Pantone colors not available).
If a single channel, then the stream channel will be shown in a blue single-line symbol (line weight not available).
All islands resolvable on the source document will be shown.

1:62 500

Delimiting rules as for 1:24 000.

1:100 000

If stream is compiled from source, follow Data Extraction Guidelines (above).
If generalized from larger scale, then
aggregate islands and small bars > 0.02 inches in both directions by convex hull.
eliminate islands and small bars < 0.02 inches in both directions.
eliminate small lakes and depressions < 0.02 inches in both directions.
simplify remaining channel(s).

1:250 000

Delimiting rules as for 1:24 000.

1:500 000

Delimiting rules as for 1:24 000.

Special conditions at all scales (specific definition rules for this feature instance)
The stream representation will change from the linear (single-line symbol) to the areal (double-line symbol with tint fill) representation when the channel width becomes greater than 0.02 inches. To accommodate variations in stream width:
if 0.01 inches < stream width < 0.02 inches for longer than 2.64 inches
then represent stream channel as an area, otherwise as a line.
if 0.02 inches < stream width < 0.03 inches for less than 2.64 inches
then represent stream channel as a line, otherwise as an area.
Transitional tapering from a linear representation to areal representation will take place over a distance of 0.15 inches prior to the 0.02 inch width requirement, and vice versa.

7.5 Summary

Ideally, a cartographic data model linking multiple representations for all scales at which generalization might occur should embed knowledge about how to effect variations in the cartographic (surface) meaning to preserve associations with geographic (deep) meaning. The case studies demonstrate how object-oriented encapsulation can embed rules to automate these associations. The rules guiding surface variation include not only representational encapsulations, but also encapsulations for data extraction and delineation. This a curious discovery. Conventional wisdom would lead one to believe that representational rules should guide the form of the desired display (the surface meaning). The case study work reported here indicates that the rules by which data are collected and defined will impact the representation regardless of other rules put in place.

As suggested at the outset of the case studies, it is evident that scale-dependent variations in geometry and topology tend to occur with scale change. For the industrial complex, catastrophic change is apparent in between 1:62 500 and 1:100 000. In this range, all buildings are eliminated, as is the feature label. Transportation elements (roads and railway) are retained. As these might be categorized in a different feature domain (transportation as opposed to industrial complex) it is difficult to say if the representation rules at 1:100 000 should dictate that the complex is not shown, or that its depiction should be limited to transportation-related features.

For the stream, catastrophic change on the map seems most apparent between 1:100 000 and 1:250 000. However, in the data model, the representational rules do NOT change within this range — the difference in depiction is literally caused by the Special Conditions, specifically the high frequency of details that fail to meet the size threshold at the reduced scale. Thus, the knowledge embedded in the data model rules does not necessarily indicate at which scale one might expect catastrophic changes to occur. That is, surface changes are conditional on the deep meaning, or geographical content of the data. This is a key point.

It is interesting to note ambiguities that exist regardless of the specificity of representational rules. For example, the rule that stream channel names should 'follow the main channel' will be more easily followed in some but not all locations along the channel, and this decision may require manual intervention or at least special pattern recognition routines. Both are examples justifying application of amplified intelligence techniques to map generalization (Weibel, 1991).

To summarize, any single data modeling case study is not the same as the real-world phenomenon; it is an abstraction of the phenomenon suitable to a particular scale and purpose. One should keep in mind that two case studies from a single set of map sheets cannot by themselves justify adoption of a data model extension. This paper presents case studies as a proof-of-concept, as a stated intention to continue this research. A complete inventory of all DLG-E features at all possible scales is not realistic. However, representative samples of features sampled from topographic sheets compiled by various NMAs may prove informative for international comparisons of data modeling experiences.

Impediments to implementation beyond the prototype level described in this paper include formalizing knowledge about what ranges of scale are appropriate for each representation, and what role should be played by geographical context in rule formation. The selection of compilation scales in the DLG-E data model (as with any NMA) is arbitrary, and propagates by convention into general usage. More research is needed to identify what are the critical scales at which surface (cartographic) meaning must be modified to preserve its association with deep (geographic) meaning. There is some evidence in these case studies that some, but not all, feature components change with scale. In the aluminum plant case study, for example, transportation routes are preserved long after storage buildings have been deleted. The implication is that deep meaning associations may be preserved using polymorphic procedures that affect some but not all feature components at specific scales. Additional information on this aspect of scale-dependent data modeling can be provided empirically, by modeling some of the scale changes using only the source features (at 1:24 000) and the stated size thresholds, and then comparing the resulting generalized features with compiled map features. This thread will inform researchers about the biggest impediment, that of formalizing knowledge about what ranges of scale are appropriate for each set of representation rules, and what role should be played by deep geographical meaning in the formation of surface representational rules.

Acknowledgements

This research forms a portion of NCGIA Research Initiative 8, 'Formalizing Cartographic Knowledge', funded by the National Center for Geographic Information and Analysis (NCGIA). Support by the National Science Foundation (NSF Grant SES 88-10917) is gratefully acknowledged. This research was completed while the author was a Visiting Research Scholar, in residence at the US Geological Survey, Reston, VA. Sabbatical support by USGS National Mapping Division is also acknowledged.

References

Adeli, H., 1990, *Knowledge Engineering, Volume 1, Fundamentals*, New York: McGraw-Hill.

Borgida, A., 1986, Conceptual modeling of information systems, in Brodie, M.J. and Mylopoulos, J. (Eds) *On Knowledge-Based Management Systems*, pp. 461–9, New York: Springer-Verlag.

Brassel, K.E. and Weibel, R., 1988, A review and conceptual framework of automated map generalization, *International Journal of Geographical Information Systems*, **2**(3), 229–44.

Bruegger, B.P. and Frank, A.U., 1989, Hierarchies over topological data structures, *Proceedings of the ASPRS/ACSM Annual Convention*, Baltimore, MD, Vol. 4, pp. 137–45.

Buttenfield, B.I., 1989, Scale dependence and self-similarity of cartographic lines, *Cartographica*, **26**(1), 79–100.

Buttenfield, B.P., 1991, A rule for describing line feature geometry, in Buttenfield, B.P. and McMaster, R.B. (Eds) *Map Generalization: Making Rules for Knowledge Representation*, pp. 172–86, London: Longman.

Buttenfield, B.P., Weber, C.R., Leitner, M., Phelan, J.J., Rasmussen, D.M. and Wright, G.R., 1991, How does a cartographic object behave? Computer inventory of topographic maps, in *Proceedings of GIS/LIS '91*, Atlanta, Georgia, November, Vol. 2, pp. 891–900.

Corbett, J.P., *Topological Principles in Cartography*, Washington, DC: US Department of Commerce Technical Paper Number 48.

FIPS, 1992, *The Spatial Data Transfer Standard*, Washington, DC: Federal Information Processing Standard Number 173.

Guptill, S.C. (Ed.), 1990, *An Enhanced Digital Line Graph Design: A Feature-Based Model for Digital Spatial Data Bases that Represent Geographic Phenomena*, US Geological Survey Circular 1048, Washington, DC: US Government Printing Office.

João, E., Herbert, G., Rhind, D., Openshaw, S. and Raper, J., 1993, Towards a generalization machine to minimize generalization effects within a GIS, in Mather, P. (Ed.) *Geological Information Handling: Research and Application*, pp. 63–78, New York: Wiley & Sons.

Lagrange, J.P., Ruas, A. and Bender, L., 1993, 'Survey on Generalization', Paris: IGN/COGIT Internal Report DTO93-0538, 48 pp.

Leitner, M., 1993, 'Multiscale inventory of a topographic map series', unpublished MA thesis, Department of Geography, SUNY-Buffalo, NY.

McMaster, R.B. and Shea, K.S., 1992, *Generalization in Digital Cartography*, Washington, DC: Association of American Geographers Resource Monograph.

Müller, J.C., Weibel, R., Lagrange, J.P. and Salgé, F., 1993, Generalization: State of the Art and Issues, Position Paper, *GISDATA Task Force on Generalization*, Sheffield, UK: European Science Foundation Scientific Programme on GIS Data Integration and Data Base Design.

Nyerges, T., 1980, 'Modeling the structure of cartographic information for query processing', unpublished PhD dissertation, Department of Geography, Ohio State University.

Nyerges, T., 1991, Geographic information abstractions: conceptual clarity for geographic modeling, *Environment & Planning A*, **23**, 1483–99.

Openshaw, S., 1984, The modifiable areal unit problem, *CATMOG*, Paper number 38.

Parsaye, K., Chignell, M., Khosshafian, S. and Wong, H., 1989, *Intelligent Databases: Object-Oriented, Deductive, Hypermedia Technologies*, New York: John Wiley.

Peuquet, D., 1984, A conceptual framework and comparison of spatial data models, *Cartographica*, **21**(4), 66–113.

Richardson, L.F., 1961, The problem of contiguity: an appendix to the statistics of deadly quarrels, *General Systems Yearbook*, **6**, 139–87.

Richardson, D.E., 1988, Database design considerations for rule-based map feature selection, *ITC Journal*, **2**, 165–71.

Richardson, D.E., 1989, Rule-based generalization for base map production, in *Proceedings, Challenge for the 1990s: Geographic Information Systems Conference*, Canadian Institute for Surveying and Mapping, Ottawa, Canada, pp. 718–39.

Richardson, D.E. and Müller, J.C., 1991, Rule selection for small scale map generalization, in Buttenfield, B.P. and McMaster, R.B. (Eds) *Map Generalization: Making Rules for Knowledge Representation*, pp. 136–49, London: Longman.

USGS, 1993, 'Statement of Mission', Internal Document, Washington, DC: US Geological Survey.

Weibel, R., 1991, Amplified intelligence and rule-based systems, in Buttenfield, B.P. and McMaster, R.B. (Eds) *Map Generalization: Making Rules for Knowledge Representation*, pp. 172–86, London: Longman.

8

Holistic generalization of large-scale cartographic data

Geraint Ll. Bundy, Chris B. Jones and Edmund Furse

Department of Computer Studies, Pryfysgol Morgannwg/University of Glamorgan, Mid Glamorgan, CF37 1DL, UK

8.1 Introduction

Cartographic generalization can be described as the abstraction and simplification of items on a map according to their relative importance and contribution to the intended purpose of the map. Generalization is context based in that the actions to be applied to any feature are governed not just by the feature itself but by the surrounding features and the cartographic imperative involved. This context dependence has been described by many researchers, e.g. Raisz (1962) and Robinson *et al.* (1978).

This research involves the implementation of a prototype system called MAGE (Map Authoring and Generalizing Expert) that can perform a representative sample of cartographic procedures. The MAGE system is implemented on a Sun 4 workstation using the Kappa expert system development tool (Intellicorp, 1993). Kappa is an object-oriented tool that allows deterministic procedures to be implemented using C (or C++) and non-deterministic procedures and rules (with backward and forward chaining) to be expressed in a dedicated language called ProTalk. MAGE is designed to be a test bed suitable for the incremental development of a practical, albeit experimental, generalization system. It is envisaged that development work would be performed by an experienced cartographer and programmer team and that everyday operation of MAGE would require only a little experience.

The approach used in MAGE is founded on two key ideas. First, that a topographic map surface can be well represented by a structure based on simplicial complexes which facilitates some of the basic generalization operations. Second, that the semantic structures needed for generalization can be represented by a hierarchy of context frames, each of which encapsulates the knowledge required to recognize, generalize and resolve a cartographic situation.

A topological data structure called the simplicial data structure (SDS) has been implemented. Each map object in the SDS is represented by a set of 2-d simplices (triangles) that are maintained in the form of a constrained Delaunay triangulation (Chew, 1987; De Floriani and Puppo, 1988; Aurenhammer, 1991; Beard, 1991). This structure gives a fully connected 2-d plenum that implicitly stores object-level topology. Useful relationships such as 'adjacent' and 'between' are evident in the sets of simplices connecting objects. The SDS can be used directly to perform several types of operations necessary for automatic generalization, e.g. automatic overlap detection, displacement, merging, enlargement and skeletonization.

A frame-based representation is proposed which provides a framework for knowledge in various forms and for various purposes (Minsky, 1981; Armstrong, 1991). Structural knowledge is stored in recognizer agents which are able to identify structures and contexts in the map data. Operational knowledge is held in generalizer agents which know how to apply a cartographic transformation to a map object or a structure of such objects. Resolving agents are applied to conflict situations that might be caused by the actions of generalizers. Context frames coordinate these agents and provide working storage. Rule bases control the initiation of context frames and the application of their agents.

MAGE attempts to generalize a map holistically, that is, by considering all features on the map and the interaction between them. This is much more ambitious than most previous attempts at automation which deal only with certain features in isolation, e.g. Lichtner (1979), Chrisman (1983), Lay and Weber (1983), and Meyer (1986). It is clear that humans generalize holistically — they may focus on a particular subset of map features while performing generalization but they always take into account the context in which these objects are situated.

The problem with generalizing map features independently is that situations where disparate classes should interact during a generalization process cannot be dealt with. Consider the following simple example. It may happen for a certain style of map that some point features (perhaps a trigonometric point) have a higher priority for spatial accuracy than a linear feature (say, a road). If the road is smoothed without regard for that point then the road may come to lie over the trigonometric point or even on the other side of the point. This kind of change in topology is undesirable and should be avoided, if possible. A topological inconsistency can be detected and then corrected but it would be better to use the knowledge that a trigonometric point has a higher priority than a road to direct the generalization in the first place.

The following sections describe:

- the SDS and some of the major operators implemented on it,
- the knowledge representation used in MAGE, and
- discussion and conclusion.

8.2 The simplicial data structure

It has been recognized that displacement and merging are major problems during the generalization process (Chapter 1). A topological data structure that facilitates these operators and overcomes many of the problems of an implementation based on Euclidean geometry (Egenhofer *et al.*, 1990) forms a useful representation for the purposes of automated map generalization. A map surface can be represented by a set of simplicial complexes. Each object on the map consists of a number of non-overlapping simplices. The space between objects is also represented by simplices. For example, see Figure 8.1.

Such a representation can be viewed as a 2-d plenum, i.e. a filled space. No point on the map is unattributed or unconnected. Since the space is fully connected, changes in the

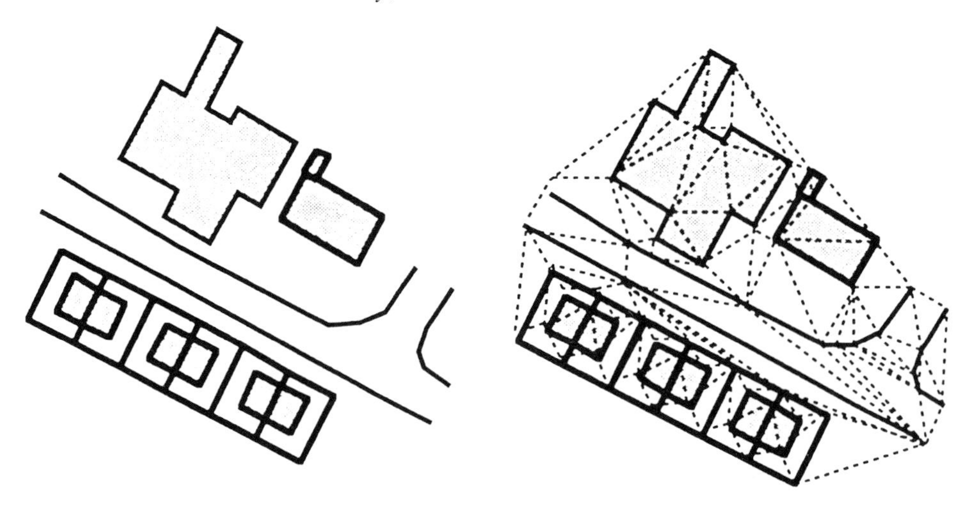

Figure 8.1 A portion of a map and the SDS representation.

space can be propagated relatively easily. The SDS stores both metric and topological information which avoids inconsistencies when manipulating the map. The SDS does not preclude higher level spatial indices being applied to the data if they are necessary or useful, e.g. hierarchical structures similar to quadtrees (Bruegger, 1989).

8.2.1 Simplices and simplicial complexes

A simplex is the simplest k-d polyhedron, i.e. point for 0-d space, line for 1-d, triangle for 2-d, tetrahedron for 3-d, and so on. A simplex of dimension k is composed of $k+1$ simplices of dimension $k-1$. A simplicial complex is a finite collection of simplices and their faces (Jackson, 1989; Egenhofer *et al.*, 1990). A simplicial complex can be used to model a cartographic object and a collection of complexes make up a map.

Let O be the set of all objects, T be the set of all triangles, E be the set of all edges and V be the set of all vertices. Then we define the following:

- **map**: a finite set of objects. $M = (O_0, O_1, O_2, \ldots O_m)$
- **object**: a simplicial complex. This is a finite set of 2-simplices (i.e. triangles). $O=(T_0,T_1,T_2, \ldots T_n)$
- **triangle**: the triangle is the 2-d simplex and consists of three 1-d simplices (edges). $T=(E_1, E_2, E_3)$ where $E_i \in E$
- **edge**: the edge is the 1-d simplex and consists of two 0-d simplices (points). Edges can only join at vertices. $E = (V_1,V_2)$
- **vertex**: the vertex is the 0-d simplex and consists of an x,y coordinate pair in a plane cartesian coordinate system. $V = (x, y)$ as $x,y \in \mathbb{R}$.

8.2.2 Implementation

The SDS is implemented as linked lists and symbolic lists. Objects are defined in the Kappa environment. Triangles, edges and vertices are stored as doubly linked lists of appropriate C structures. The object to triangle relationship is held as a Kappa list of pointers to triangle structures. The other relationships: triangle–edge, edge–vertex and

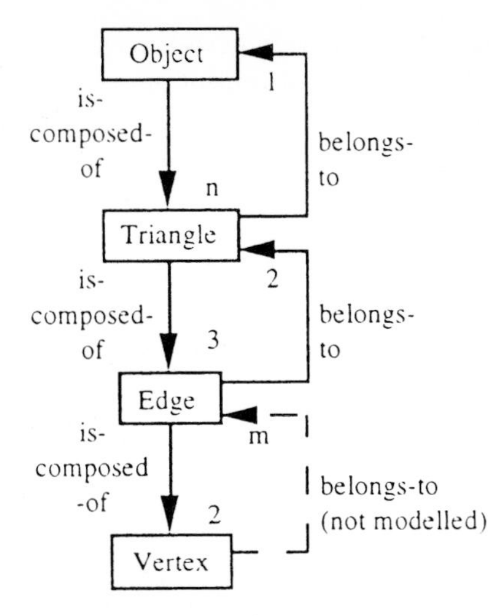

Figure 8.2 Entity–relationship diagram for SDS.

triangle–triangle are represented by pointers to the appropriate nodes in the linked lists. See Figure 8.2. Triangles that belong to no particular cartographic object are termed *free-space triangles.*

8.2.3 Automatic detection and resolution of spatial conflicts

Definition of spatial conflict

Spatial or graphical conflict can be defined as a situation where more than one object covers any part of the space enclosed by the map, i.e. where a triangle or a part of a triangle on the map is inside another triangle. Such conflicts can be detected efficiently and the SDS provides valuable information for their resolution because the extent and direction of a conflict can be deduced from the set of inverted triangles indicating the conflict.

Detecting conflict

The detection of spatial conflict can be performed by comparing the actual geometry of the objects on the map with the stored topology in the SDS. The technique is based on the fact that if the order of the vertices is changed (this is called an *inversion*) after an operation, then a point (at least) has crossed an edge. This might just mean that a free-space triangle has been overlapped, in which case we can simply retriangulate the affected area. If, on the other hand, a cartographic object's triangle has been overlapped then resolving action must be taken.

Resolution of conflict

There are several ways to deal with a conflict condition once it has been detected. The way in which it is handled will depend on the particular situation that the conflict represents and the confidence of the recognition system classifying the situation. Possible resolvers for a spatial conflict are: (1) displacement of one or all of the objects involved, (2) omission of one or some of the objects, (3) merging the objects, and (4) changing the shape of one or some of the objects by smoothing or 'squashing'.

The set of inverted triangles caused by a conflict indicates the extent of the conflict, the nature of the conflict (whether a whole object, single corner, or perhaps two separate parts of an object are involved), and the direction of the overlap.

8.3 Major operators on the SDS

8.3.1 Object scaling

Objects are scaled during generalization for a variety of reasons. Objects are enlarged if they are too important to omit but too small to display at the target scale. Objects may need to be scaled down in size during a typification process or after simplification (e.g. to maintain a constant area).

Scaling can be achieved by applying a *vector of displacement*, **R**, to each point in the object outline in turn. The scaling is specified as a buffer width, δ, which, when positive, produces an expansion in the object outline and, when negative, produces a contraction in the object outline. The vector of displacement is calculated for each point in the object outline as the vector sum of the normals (in the outward direction and of length δ) of the two edges flanking the point p (p_1 and p_2). For orthogonal objects, i.e. where successive edges in the outline are at right angles with respect to each other, this achieves scaling very efficiently using one vector addition per point in the outline: see Figure 8.3. The same technique can be used for non-orthogonal objects but a correction must be applied to the vector of displacement. Conflicts can occur with other objects or with other edges of the same object (e.g. in a concavity in the outline) and must be dealt with by further processing. These conflicts are readily detectable since they cause inversions in the SDS.

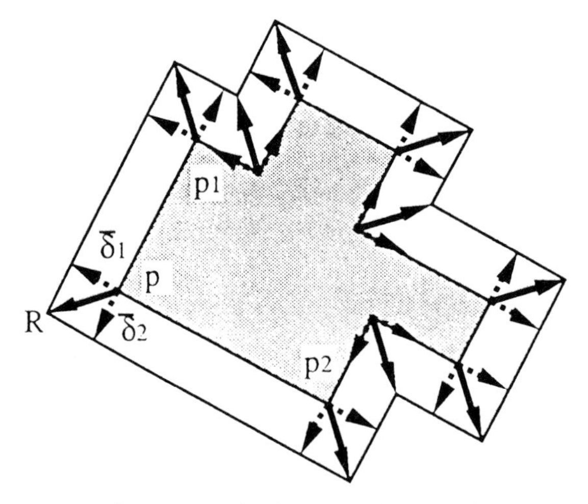

Figure 8.3 Scaling an orthogonal object.

8.3.2 Skeletonization

Skeletonization can be used for finding the 1-d analogue for an areal object and is also useful during merge and displacement operators. The SDS allows for an approximation of the medial axis transformation to be calculated very efficiently. The approximation is based on the lemma that the mid-point of the diagonal of a trapezoid formed by two triangles is exactly halfway between the opposite edges of the trapezoid (see Figure 8.4). This is founded on work by Chithambaram *et al.* (1991). The approximation works best on objects with a relatively high density of points in their outlines.

The skeleton can be biased toward one side of the shape simply by selecting points at a proportion of the edge's length other than 50 per cent. This is particularly useful for merging and displacement where the skeleton approximation can be used to move objects together by a proportion based upon some attribute such as importance or size.

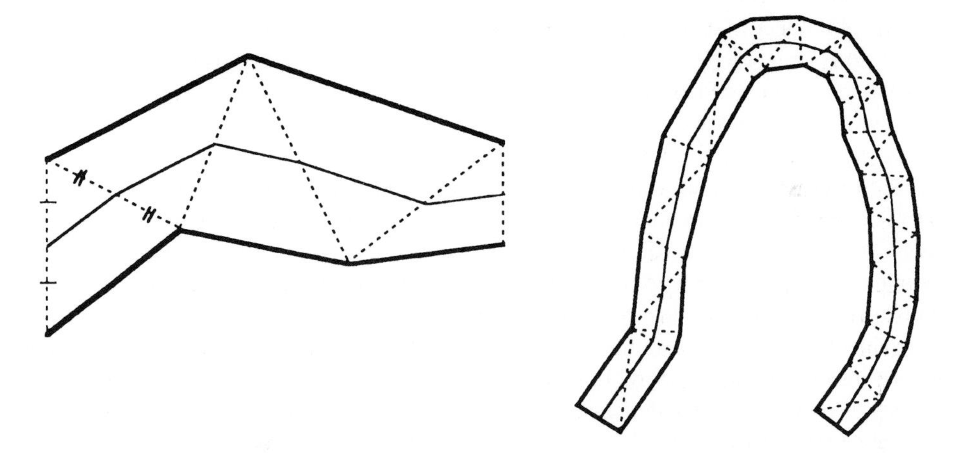

Figure 8.4 The approximation of the skeleton for a linear object.

8.3.3 Displacement

Displacement is a necessary but problematic operation in map generalization. It can be caused by many generalization actions — symbolization, simplification, smoothing, etc. Other generalization operators need controllable displacement as a tool to achieve the desired result — merging, for example.

One major difficulty is controlling the propagation of displacement in crowded map areas — this occurs when an object is displaced for some reason and then causes a conflict with a nearby object which then has to be displaced, which causes another conflict, and so on. Limiting displacement propagation, or damping, can be achieved by utilizing the free space between successive features in the displacement chain by as much as possible. However, in very crowded areas this may not be sufficient and more severe action must be taken, perhaps the removal of a feature to make way for the displacement.

The SDS can be used to guide and control displacement. Examining the simplex set around the object can reveal the amount of free space in any direction, permitting or precluding attempts at displacement in those directions. It may be beneficial to look further ahead when considering displacement by estimating the probability of causing propagation in nearby features.

Examples of how the SDS can be used for displacement are given in the next section.

8.3.4 Merging objects

The graphical combination of map objects through merging is an essential generalization operator. There are various flavours of merge operation that can be implemented — each kind uses the information held in the simplices between the objects to be merged in a different way. The proportion of displacement for each object involved can also be varied so that, perhaps, small objects gravitate toward larger objects; less 'important' toward more 'important' (in merges involving the reclassification of different classes of features); objects could meet mid-way or not be displaced at all.

Two types of merge operators have been identified for use on orthogonal objects such as buildings. Their use depends upon a reasonably small number of triangles between the two objects being merged. Such operators would be applied iteratively to a cluster of objects so that only two objects were being dealt with at any one time.

The direct-merge operator maintains the alignment of the objects by moving the objects together directly, i.e. only facing edges are aligned. The relationship between the facing edges is represented by the triangles that have an edge belonging to one object and a point belonging to the other. The altitudes of these triangles could be used to deduce the adjustment needed to make the facing edges parallel and the displacement needed to bring them together. For example in Figure 8.5 (a) and (b), the small building is being merged into the larger building. In this case, the vector of displacement is taken from the altitude of one of the triangles separating the two buildings. Once the edges are aligned internal edges can be removed ((c) and (d)).

The snap-merge operator (Figure 8.6) attempts to align the objects' nearest parallel edges. This can be achieved by aligning the merge vector with the shortest outer connecting simplex edge.

When merging natural objects, a freer approach can be taken — there is no need to align the objects. Plastic merging allows objects to be moulded together forming a smooth boundary, as in Figures 8.7 and 8.8, and makes use of the skeleton approximation discussed in section 8.3.2. Plastic merging is achieved by combining the edges and points in the outlines of the two objects that are connected by a triangle. A point in the outline

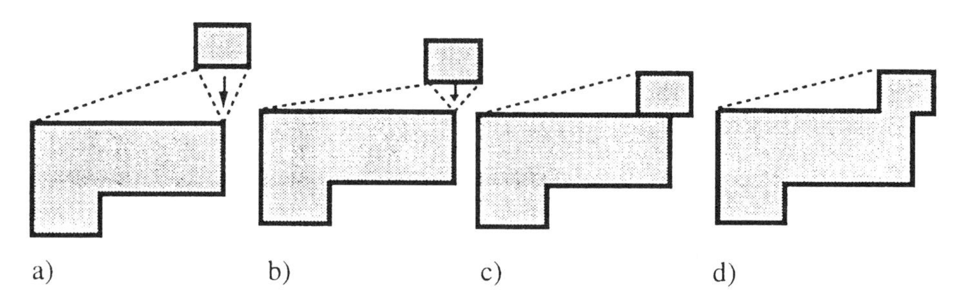

Figure 8.5 Direct building merge.

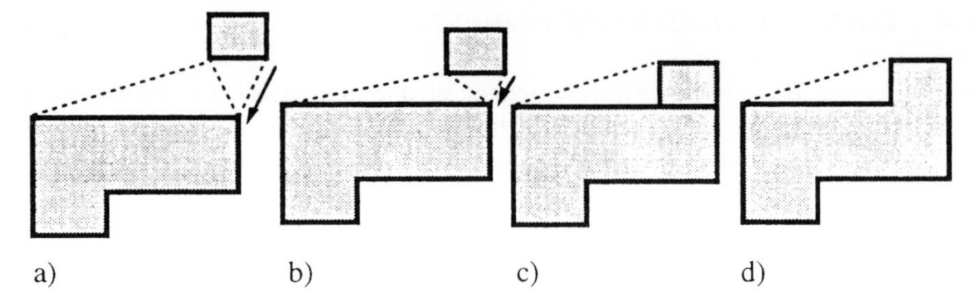

Figure 8.6 Snap merge.

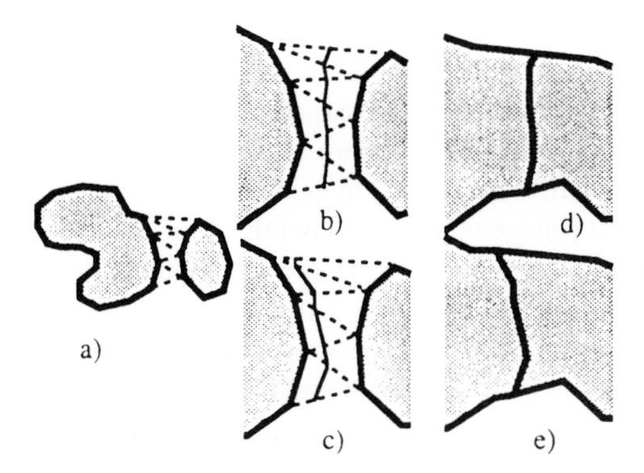

Figure 8.7 Plastic merge with no displacement ((b) and (d) 50 per cent merge line, (c) and (e) 75 per cent merge line).

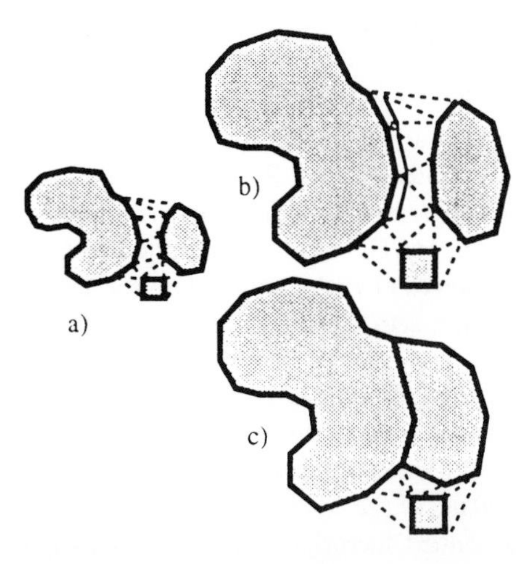

Figure 8.8 Plastic merge with displacement (80 per cent merge line).

of one object that is connected to an edge in the other is split to form a new edge. These two edges are combined at the location of the associated skeletal edge. Displacement can be used to bring the centres of the objects closer together and to minimize the increase in area caused by merging (Figure 8.8).

8.4 Generalization using contexts

If 'a map is an image that serves as a medium of communication' (Freeman, 1991) then map generalization can be viewed as an interpretation process and likened to natural language interpretation. When interpreting natural language, humans probably perform the following processes:

- recognition: matches an input with an internal concept;
- understanding: places the received concept in the correct context (i.e. with respect to previous information);
- translation: converts the internal concept(s) into a new form.

When receiving a sentence we often *expect* the next word, phrase or sentence. This is because our recognition of the concept is guided by the context set up in the previous sentences. Sometimes our expectations are incorrect and, as other sentences are received, we have to revise our internal model of what we are hearing. It is usually not possible to understand fully a sentence without taking it in context. The same is true of those entities on a map that are dealt with in cartographic generalization.

The recognize–understand–translate process can be applied to cartographic generalization. The task of generalization can be expressed as a series of expected contexts, i.e. typical situations on a map that require generalization, where the context is spatial (topological relationships), semantic (the meaning of the objects) and attributive (the kinds of objects). In MAGE, the topological relationships are stored in the SDS and the attribute and semantic knowledge are stored in frames and objects. Representing generalization as a set of contexts requires an analysis of the contexts that can be isolated within the map, recognized by the system, and finally generalized. It imposes a degree of objectivity onto the specification of the generalization process that is not usually found in specifications for human cartographers.

8.4.1 Contexts

A context is represented by a context frame. Recognizing agents form the link between the concept of a particular cartographic situation and the knowledge about what can be done to deal with that situation. MAGE has to be supplied with a recognition technique that can uniquely identify instances of each context. Once a context has been identified, MAGE can set about applying a set of generalization operators to achieve the desired transformation.

Recognizers are written by the map author in increasing specialization as the context classification itself becomes more specialized. The system can be implemented initially with a small number of context frames, recognizers and generalization operators and, because of the independent nature of the context frames, can be developed by adding new contexts. An example of a context hierarchy can be seen in Figure 8.9. Note that the name of a context indicates the situation that it represents not the operation that might be applied to resolve it.

An example of a context frame might be 'objects on the map that are too small to be included' (`ObjectTooSmall`). A general context frame would be written that searches the map for objects that are smaller than a certain minimum area and removes them. However, the map requirements might call for a more specific action to occur under certain conditions, perhaps a small building would be merged with a nearby building rather than being removed (`BuildingMerge`) or a building in a remote area

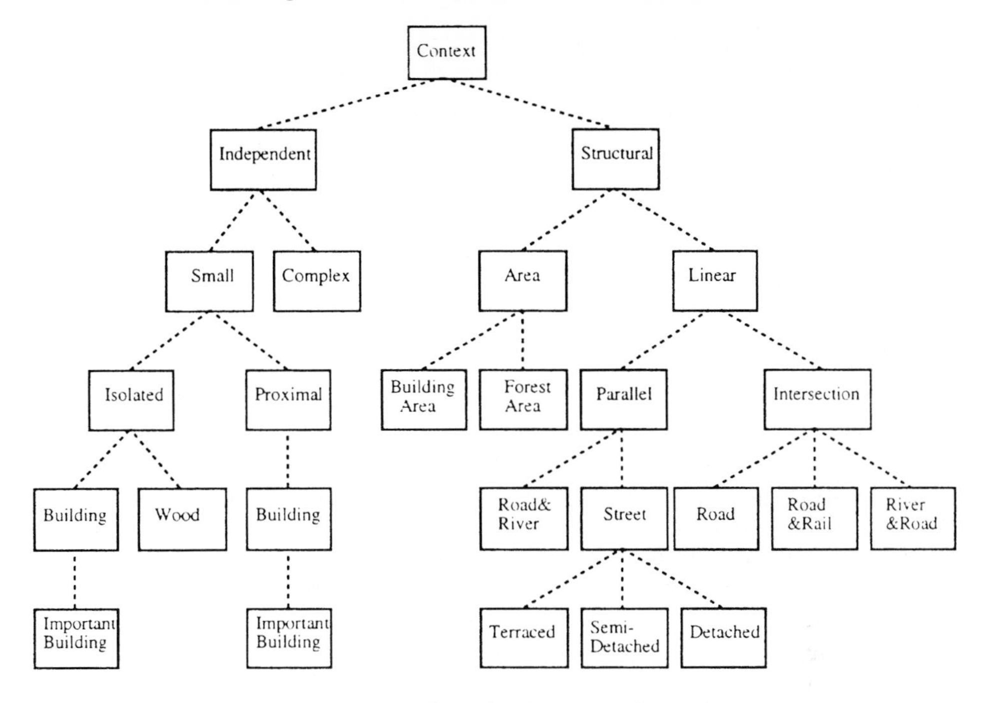

Figure 8.9 An example of a context hierarchy.

(`BuildingIsolated`) would be retained because it forms an important landmark. These situations would be represented by context frames descended from the `ObjectTooSmall` frame.

One object might satisfy a number of these recognition processes. Mediation between frames competing for objects can be achieved by using a confidence rating for the recognizer. However, many recognizers will give a binary confidence rating, i.e. either it will be 100 per cent sure or will not recognize anything: so more criteria are needed. This might be addressed by adopting a maximum specificity priority rule and then an arbitrary priority rating to resolve competition between equally specific frames. For example, if an object is held in two frames, say, `ObjectTooSmall` and `BuildingIsolated`, then the more specific of the two frames wins the object and gets a chance to generalize it first. If, on the other hand, the two frames competing for the object are of the same specificity, then some priority given by the map author and stored in the frame could be used. Obviously other strategies are possible, perhaps based on the size of the structures involved in each context frame — giving priority to the context with the largest number of objects would mean that more generalization could be achieved as soon as possible. The current experimental approach, which is discussed in the next section, is simply to deal with any context as soon as it is recognized.

A context consists of a unique recognizer, a number of attribute slots, some stuctural slots and one or more generalizing agents, e.g.:

<u>SemiDetachedStreetFrame:</u>

```
BaseMass:         0.0
BasePriority:     0.0
StructMass:       ?
StructPriority:   ?
```

```
Confidence:         ?
Description:        Holds knowledge about semi-detached streets
SitStruct:          ?
MergeHouses!:       BuildingMerge!
Recognizer!:        SemiDetachedStreetFrameRecognizer!
SymbGardens!:       SymbolizeGardens!
MergeWithRoad!:     MergeLinearObjects!
Generalizers:       SemiDetachedStreetGeneralizer!
```

The `BaseMass`, `StructMass`, `BasePriority` and `StructPriority` slots are used to represent structural masses (related to inertia, i.e. reluctancy be displaced) and priorities (arbitrary importance) when a context structure is constructed. Each object in the structure contributes a Mass or Priority increment to the structure. These slots can be used to model the principles that large objects are more important and less likely to move. The question marks (?) in the slots means that the value is undefined; these slots would be filled in during instantiation. Slots and slot values with exclamation marks (!) after their names are methods or method slots. These are programs or rule bases that perform some action or find out some fact.

8.4.2 Recognizers

A recognizer is an agent (i.e. a Kappa object with procedural or declarative knowledge) that can uniquely identify instances of a context in a map. For example, the Ordnance Survey specify the generalization of streets according to their type: terraced, semi-detached or detached. In order to differentiate between these types of street, the map author must define a different context frame for each type: the generalization is different, the structural representation is different, and the recognizer is different.

The semi-detached street recognizer deals with instances of streets that roughly match the pattern: *garden, house, house, garden, garden, house, house, garden, garden,* etc. whereas the detached street recognizer looks for streets that match the pattern: *garden, house, garden, garden, house, garden,* etc. For example, see Figure 8.10.

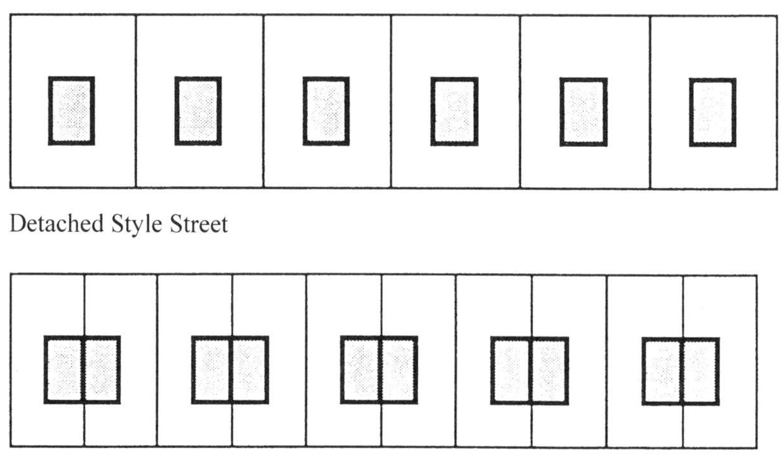

Figure 8.10 Examples of semi-detached and detached style streets.

The Kappa inheritance mechanism means that duplication is minimized in the context frames and only those elements that are actually different need to be redefined.

8.5 Control and management of generalization

The MAGE system can be thought of as a contextually organized production rule system where rules have antecedents (recognizers) and consequents (generalizers) which both contain procedural and/or declarative knowledge. The recognizers and generalizers operate on a mapspace where topology is stored in simplicial complexes, and attribute and semantic information is stored in context frames and objects. Ordinary rule-conflict resolution may not be adequate for the control of such a system because of the relatively large overheads, in terms of both processing time and memory space, incurred in the matching process.

An investigation is currently underway into the issue of control in MAGE but a relatively straightforward approach would be the use of a generalization agenda and an immediate activation of contexts as they are recognized (as opposed to finding many contexts and having to decide which one to activate). The generalization agenda describes the major steps in the generalization process. It consists of a list of context frames that are activated in turn. Once active, a context searches through the map for examples of itself. When it finds one, it creates an instance frame and fills in the details in the frame with actual values from the map. If the generalization of a context fails (perhaps because it causes too much spatial conflict) then the context would be deferred until a later time. Operations near that context could reactivate the context and, if the situation still matched the recognizer, generalization could be attempted again.

8.6 Discussion

MAGE is designed to generalize a map holistically; that is, it deals with many different types of map object and the interaction between them — it does not separate the map up into layers containing only one type of feature and then attempt to recombine them after generalization.

The simplicial data structure offers distinct advantages to a map generalization system — it provides an explicit topological representation that overcomes the problems inherent in Euclidean geometry (Egenhofer *et al.*, 1990) and the relationships between adjacent objects are easily identifiable. It can be used directly for many of the operations needed for generalization and allows the automatic detection of spatial conflict. The current implementation of the SDS is highly dynamic and fast but the use of a persistent, object-oriented topological database would provide greater security and flexibility and allow much larger maps to be processed.

Context frames provide a novel way of looking at the problem of representing conceptual knowledge for map generalization. The object-oriented nature of these structures gives all the benefits associated with that paradigm, e.g. encapsulation, inheritance, clustering, and extensibility (Kemp, 1990). They offer a useful and natural formalization for the causes and processes of generalization.

At the time of writing this chapter, the MAGE system is still under development. The SDS is implemented and some of the operators described in section 8.3 have been implemented at least in prototype form. The context frame hierarchy has been defined and some recognizers written. The simple control strategy described in section 8.5 has been implemented but not applied to real map data.

There is a great deal of scope for research both into operators on the SDS and control mechanisms applicable to context frames. The implementation of a backtracking mechanism would allow deeper searches into the generalization search space. The approach lends itself to parallelization — generalization is a process that needs plenty of processing power. The recognizer agents, in particular, would seem to be ideal candidates for parallel processing.

8.7 Conclusion

The paper describes a novel approach to automatic map generalization using the topological objects called simplicial complexes and a paradigm for the representation of semantic knowledge and control based around the idea of context frames.

The implementation, called MAGE (Map Authoring and Generalizing Expert), has demonstrated that simplicial complexes can be used for several fundamental operators necessary for generalization and is particularly useful for deriving the topological relationships between nearby objects.

The knowledge representation and control mechanism should allow a map author to develop a system that is capable of automatically generalizing a large proportion of a map. Context frames cluster the knowledge pertaining to a generalization problem, giving a new way of describing generalization and allowing incremental development and testing.

Acknowledgements

The authors are grateful to the Ordnance Survey for their data and friendly advice and to J. Mark Ware for the use of his constrained Delaunay triangulation program. The project is funded by a Science and Engineering Research Council studentship in collaboration with the Ordnance Survey awarded to Geraint Bundy.

References

Armstrong, M.P., 1991, Knowledge classification and organization, in Buttenfield, B.P. and McMaster, R.B. (Eds) *Map Generalization: Making Rules for Knowledge Representation*, pp. 86–102, London: Longman.

Aurenhammer, F., 1991, Voronoi diagrams — a survey of a fundamental geometric data structure, *ACM Computing Surveys*, **23**(3), 345–405.

Beard, K., 1991, Generalization operations and supporting structures, *Technical Papers, ACSM-ASPRS Annual Convention, Auto-Carto 10*, Vol. 6, pp. 29–45.

Bruegger, B., 1989, Hierarchies over topological data structures, *Technical Papers, ASPRS-ACSM Annual Convention*, Baltimore, MD, Vol. 4, March, pp. 137–45.

Chew, P.L., 1987, Constrained Delaunay triangulation, in *Proceedings of the Third Symposium on Computational Geometry*, ACM, pp. 215–22.

Chithambaram, R., Beard, K. and Barrera, R., 1991, Skeletonizing polygons for map generalization, *Technical Papers, ACSM-ASPRS Convention*, Cartography and GIS/LIS, Baltimore, Vol. 2, pp. 44–55.

Chrisman, N.R., Epsilon filtering: a technique for automated scale changing, *Technical Papers, 43rd Annual American Congress on Surveying and Mapping (ACSM) Meeting*, Washington, DC, March, pp. 322–31.

De Floriano, L. and Puppo, E., 1988, Constrained Delaunay triangulation for multiresolution surface description, in *Proceedings of the 9th International Conference on Pattern Recognition*, Rome, Italy, 14–17 November, pp. 566–9.

Egenhofer, M., Frank, A. and Jackson, J.A., 1990, Topological data model for spatial databases, in Buchmann, A., Gunther, O., Smith, T. and Wang, Y. (Eds) *Symposium on the Design and Implementation of Large Spatial Databases*, Lecture Notes in Computer Science, Vol. 409, Zurich, Switzerland, July, pp. 814–19.

Freeman, H., 1991, Foreword, in Buttenfield, B.P. and McMaster, R.B. (Eds) *Map Generalization: Making Rules for Knowledge Representation*, pp. xi–xii, London: Longman.

Intellicorp, Inc., 1993, *Kappa Manual Set*, Intellicorp, Inc., 1975 El Camino Real West, Mountain View, CA 94040–2216.

Jackson, J., 1989, Algorithms for triangular irregular networks based on simplicial complex theory, in *Proceedings of ASPRS-ACSM Annual Convention*, Baltimore, MD, Vol. 4, March, pp. 131–6.

Kemp, Z., 1990, An object-oriented model for spatial data, in *Proceedings of the 4th International Symposium on Spatial Data Handling*, Zurich, Switzerland, pp. 659–68.

Lay, V.H.-G. and Weber, W., 1983, Waldgeneralisierung durch digitale Rasterdatenverarbeitung, *Nachrichten as dem Karten- und Vermessengswessen, Series 1*, **92**, 61–71.

Lichtner, W., 1979, Computer-assisted processes of cartographic generalization in topographic maps, *Geo-Processing,* **1**, 183–99.

Meyer, U., 1986, Software developments for computer-assisted generalization, in *Proceedings AutoCarto*, London, Vol. 2, pp. 247–56.

Minsky, M., 1981, A framework for representing knowledge, in Haugeland, J. (Ed.) *Mind Design*, pp. 95–128, Cambridge, MA: MIT Press.

OS (Ordnance Survey), 1988, *Module 7 Section E 1:10000 Series Maps — Drawing Instructions and Specifications*, Amendment No. 3, April.

Raisz, E., 1962, *Principles of Cartography*, New York: John Wiley.

Robinson, A.H., Sale, R.D. and Morrison, J.L., 1978, *Elements of Cartography*, 4th Edn, New York: John Wiley.

9

The GAP-tree, an approach to 'on-the-fly' map generalization of an area partitioning

Peter van Oosterom

TNO Physics and Electronics Laboratory, PO Box 96864, 2509 JG The Hague, The Netherlands

9.1 Introduction

The concept of '*on-the-fly*' map generalization is very different from the implementation approaches described in the paper by Müller *et al.*: *batch* and *interactive* generalization (Chapter 1). The term batch generalization is used for the process in which a computer gets an input dataset and returns an output dataset using algorithms, rules, or constraints (Lagrange *et al.*, 1993) without the intervention of humans. This is in contrast to interactive generalization ('amplified intelligence') in which the user interacts with the computer to produce a generalized map.

On-the-fly map generalization does not produce a second dataset, as this would introduce *redundant* data. It tries to create a temporary generalization, e.g. to be displayed on the screen, from one detailed geographic database. The quick responses, required by the interactive users of a GIS, demand the application of specific data structures, because otherwise the generalization would be too slow for reasonable datasets. Besides being suited for map generalization, these data structures must also provide spatial properties, e.g. it must be possible to find efficiently all objects within a specified region. The name of these types of data structures is *reactive data structures* (van Oosterom, 1989, 1991, 1993).

As will be discussed in section 9.2, an area partitioning possesses some special problems when being generalized. In order to avoid gaps when not selecting small area features, a special structure is proposed: the GAP-tree. Section 9.3 describes two other reactive data structures, which will be used in combination with the new GAP-tree: the Reactive-tree and the BLG-tree. The implementation and test results are given in section 9.4, where both visual and numerical results are shown. Finally, conclusions and future work are summarized in section 9.5.

9.2 The GAP-tree

This section first describes an area partitioning as a map basis, and the problems encountered when generalizing this type of map. In section 9.2.2 the key to the solution is introduced: the area partitioning hierarchy. The creation of the structure that supports this hierarchy, the GAP-tree, is outlined in section 9.2.3. Operations on the GAP-tree are described in section 9.2.4.

9.2.1 Problems with an area partitioning

An area partitioning[1] is a very common structure used as a basis for many maps, e.g. choropleths. Two problems occur when using the following generalization techniques:

1. **Simplification:** Applying line generalization to the boundaries of the area features might result in ugly maps because two neighbouring area features may now have overlaps and/or gaps. A solution for this problem is to use a topological data structure and apply line generalization to the common edges.
2. **Selection:** Leaving out an area will produce a map with a hole which is, of course, unacceptable. No obvious solution exists for this problem.

A solution for the second (and also for the first) problem is presented here. It is based on a *novel* generalization approach called the *Area Partitioning Hierarchy*, which can be implemented efficiently in a tree structure.

9.2.2 Area partitioning hierarchy

The gap introduced by leaving out one area feature must be filled again. The best results will be obtained by filling the gap with neighbouring features. This may be easy in the case of so called 'islands': the gap will be filled with the surrounding area, but it may be more difficult in other situations.

A basic notion is the *importance* of an area feature a, which is a function of its size and type (or role): $I(a) = f(\text{size}, \text{type})$. The size could be measured by its area $A(a)$ or perimeter $P(a)$. The weight of the type of the feature depends on the application. For example, a city area may be more important than a grassland area in a given application. This could be described with the following importance function: $I(a) = A(a) \times \text{weight factor}(a)$.

By taking both the spatial relationships and the importance of an area feature into account, an *area partitioning hierarchy* is created. This hierarchy is used to decide which area is removed and also which other area will fill the gap created by the removed feature.

9.2.3 Building the GAP-tree

The polygonal area partitioning is usually stored in a topological data structure with nodes, edges, and faces. The process, described below, for producing the area partitioning hierarchy assumes such a topological data structure (Peucker and Chrisman, 1975;

[1] In an area partitioning, each point in the 2D domain belongs to exactly one of the areas (polygons): that is, there are no overlaps or gaps.

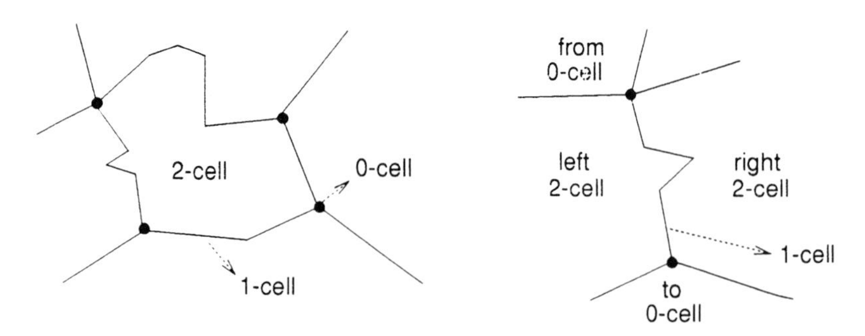

Figure 9.1 The topological data structure.

Boudriault, 1987; Molenaar, 1989; DGIWG DIGEST, 1992); see Figure 9.1. The topological elements have the following attributes and relationships.

- a **node** (or 0-cell) contains its point and a list of references to edges sorted on the angle;
- an **edge** (or 1-cell) contains its polyline, length and references to the left and to the right face;
- a **face** (or 2-cell) contains its weight factor, area, and a list of sorted and signed references to edges forming the outer boundary and possibly inner boundaries.

Note that this topological data structure contains some redundant information because this enables more efficient processing later on. After the topological data structure has been created, the following steps will produce a structure, called the *Generalized Area Partitioning* (GAP)-tree, which stores the required hierarchy.

1. For each face in the topological data structure an 'unconnected empty node in the GAP-tree' is created.
2. Remove the least important area feature a, i.e. with the lowest importance $I(a)$, from the topological data structure.
3. 'Use a topological data structure to find the neighbours of a and determine for every neighbour b the length of the common boundary $L(a,b)$.
4. Fill the gap by selecting the neighbour b with the highest value of the collapse function: $\text{Collapse}(a,b) = f(L(a,b), \text{CompatibleTypes}(a,b), \text{weight factor}(b))$. The function CompatibleTypes(a,b) determines how close the two feature types of a and b are in the feature classification hierarchy associated with the dataset (DMA, 1986; ATKIS, 1988; DGIWG DIGEST, 1992). For example, the feature types 'tundra' and 'trees' are closer together than feature types 'tundra' and 'industry' in DLMS DFAD (DGIWG DIGEST, 1992).
5. Store the polygon and other attributes of face a in its node in the GAP-tree and make a link in the tree from parent b to child a.
6. Adjust the topological data structure, importance value $I(b)$, and the length of common boundaries $L(b,c)$ for every neighbour c of the adjusted face b to the new collapsed situation.

Repeat the described steps 2–6 until all features in the topological data structure are at the required importance level (for a certain display operation). This procedure is quite expensive and probably too slow for large datasets to be performed on-the-fly. Therefore, the hierarchy is pre-computed and stored in the GAP-tree. Steps 2–6 are now repeated until only one huge area feature is left, because one cannot know what the required

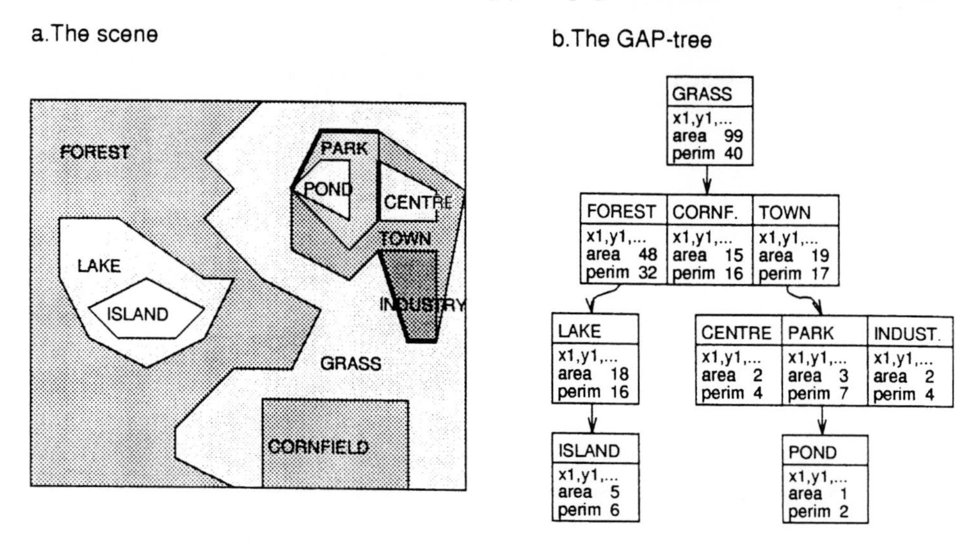

Figure 9.2 The scene and the associated GAP-tree.

importance level will be during the interactive use in a GIS. The last area feature will form the root of the GAP-tree. Furthermore, a priority queue may be used to find out efficiently which face a has the lowest importance value $I(a)$ in step 2 of the procedure.

Figure 9.2(a) shows a scene with a land-use map in the form of an area partitioning. In Figure 9.2(b) the GAP-tree, as computed by the procedure described above, is displayed. Note that a few attributes are shown in Figure 9.2(b): polygon, area, and perimeter of the final feature in the GAP-tree. The polygon is a real self-contained polygon with coordinates, and it is not a list of references. It is important to realize that this data structure is not redundant with respect to storing the common boundaries between area features. The only exception to this is the situation where a child has a common edge with another child or its parent; see the thick edges in Figure 9.2(a). More statistical information has to be obtained on how frequently this situation occurs in real datasets. As can be seen in Figure 9.2(b), the GAP-tree is a multi-way tree and not a binary tree. Some visual results of the on-the-fly generalization techniques with real data are displayed in section 9.4. The additional operations using the GAP-tree, not necessarily related to visualization, are described in section 9.2.4.

9.2.4 Operations on the GAP-tree

As a feature in the GAP-tree contains the total generalized area, the actual area A of the polygon has to be corrected for the area of its children with the following formula:

$$A(\text{actual}) = A(\text{parent}) - \sum_{\text{child} \in \text{children}} A(\text{child})$$

It is important to realize that only one level down the tree has to be visited for this operation and not the whole subtree below the parent node. In a similar way the perimeter P of a polygon can be computed, with the only difference being that the perimeters of the children have to be added to the perimeter of the parent. This results in the formula:

$$P(\text{actual}) = P(\text{parent}) + \sum_{\text{child} \in \text{children}} P(\text{child})$$

This formula for perimeter only works if the children have no edges in common with each other or with the parent. Often the boundaries of the areas in the GAP-tree are indeed non-redundant. This also enables the use of the BLG-tree for simplification of important area features on small-scale maps without producing overlaps or gaps between features. The use of the BLG-tree has a very positive effect on the response times of small-scale maps; see section 9.4.

A possible implementation difficulty is the fact that the GAP-tree is a multi-way tree. Therefore, a simple linear version has been derived from the GAP-tree by putting the features in a list based on their level in the tree. The top level feature in the tree will be the first element of this list, the second level features will follow, and so on. For example, the linear list for the scene in Figure 9.2 is: GRASS, FOREST, CORNFIELD, TOWN, LAKE, CENTRE, PARK, INDUSTRY, ISLAND, POND. When the polygons are displayed in this order, a good map can be produced without the GAP-tree. However, it is very difficult to compute the actual area without the GAP-tree.

As mentioned before, the on-the-fly generalization has been developed within the Postgres DBMS environment. Postgres is an extensible relational system. A relation does not guarantee any order among its elements. A good display can be obtained by sorting on the sequence number in the list or on the area of the feature.

9.3 Reactive data structures

There are several techniques used during the map generalization process: simplification, selection, exaggeration, displacement, symbolization and aggregation (Robinson *et al.*, 1984; Lagrange *et al.*, 1993). A reactive data structure (Projectgroep, 1982; van den Bos *et al.*, 1984) is defined as a *geometric* data structure with *detail levels*, intended to support the sessions of a user working in an interactive mode. It enables the information system to *react* promptly to user actions. This section gives a short description of tools for on-the-fly simplification and selection: the *BLG-tree* and the *Reactive-tree* are discussed in sections 9.3.1 and 9.3.2. More details can be found in van Oosterom (1989, 1991) and van Oosterom and van den Bos (1989). Note that for building the Reactive-tree, the pre-classification of the importance of a feature by a specialist is required.

Implementation of the reactive data structures in the GIS, called GEO++, has proven to be very effective (van Oosterom and Schenkelaars, 1993) especially for point, line, and individual area features: map displays based on very large datasets (> 100 Mb) can be generated within a few seconds. Note that exaggeration is used implicitly when a line (or point) feature in a small-scale map is still displayed on the screen, because the minimum line-width of one pixel at this scale is already an exaggeration of the thin line features in reality. There is still no support for the other generalization techniques: combination, symbolization, and displacement (Shea and McMaster, 1989). Van Oosterom and Schenkelaars (1993) indicate how the reactive data structures are implemented in the Postgres DBMS environment. There is a big implementation difference between the BLG-tree and the Reactive-tree, because the BLG-tree is implemented as a part of an abstract data type, while the Reactive-tree is a complete new access method. The latter is far more difficult to add because an access method interacts with many (undocumented) parts of the Postgres DBMS.

9.3.1 BLG-tree

The Binary Line Generalization tree (BLG-tree) (van Oosterom and van den Bos, 1989) is a data structure used for line generalization. Only important objects (represented by polylines) need to be displayed on a small-scale map, but without a generalization structure, these polylines are drawn with too much detail. A few well-known structures for supporting line generalization are the *Strip tree* (Ballard, 1981), the *Arc tree* (Günther, 1988), and the *Multi-scale line tree* (Jones and Abraham, 1987). In van Oosterom and van den Bos (1989) another data structure is described, the BLG-tree, which is suited for polylines, is continuous in detail level, and can be implemented with a simple binary tree.

The algorithm to create a BLG-tree is based on the Douglas–Peucker algorithm (Douglas and Peucker, 1973). Figure 9.3 illustrates the BLG-tree creation procedure and the resulting binary tree. This tree is stored for every polyline in order to avoid the expensive execution of the Douglas–Peucker algorithm each time a line generalization is needed. The BLG-tree can also be used for polygons by applying the same techniques on the boundary of the polygon.

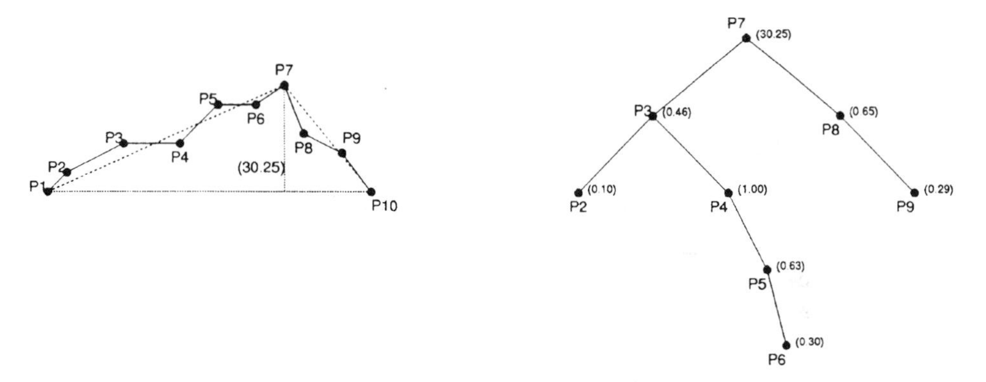

Figure 9.3 Binary Line Generalization tree.

9.3.2 Reactive-tree

The Reactive-tree (van Oosterom, 1989, 1991) is based on the R-tree (Guttman, 1984) spatial access method, and has similar properties. The main differences with the R-tree are that the internal nodes of a Reactive-tree can contain both *tree entries* and *object entries*, and the leaf nodes can occur at higher tree levels. The motivation behind this is that it must be possible to store the more important features in the higher levels of the tree; see Figure 9.4.

The further one zooms in, the more tree levels must be addressed. Roughly stated, during map generation based on a selection from the Reactive-tree, one should try to choose the required importance value such that a constant number of objects will be selected. This means that if the required region is large, only the more important objects should be selected (accessing the top levels in the tree) and if the required region is small, then the less important objects must also be selected (accessing the nodes in a certain

Figure 9.4 The scene and the rectangles of the Reactive-tree.

region of the tree). If the Reactive-tree is built well, then the total number of accesses is more or less the same. Therefore, the interaction time is also constant.

9.4 Test results

In this section the results of the performance tests of the combined use of the GAP-tree, the Reactive-tree, and the BLG-tree are presented. The DLMS DFAD[2] data (DMA, 1986) of the former Republic of Yugoslavia are used. Only the area features are used in order to evaluate the effectiveness of the GAP-tree. Section 9.4.1 shows how the Reactive-tree, BLG-tree, and GAP-tree are used in practice. Section 9.4.2 presents the benchmarks when using these data structures. The tests are performed using the GIS environment which consists of the open DBMS Postgres (Stonebraker *et al.*, 1990) and the geographic front-end GEO++ (Vijlbrief and van Oosterom, 1989) on a Sun SPARCstation II (32 Mb main memory) under SunOS 4.1.2. The test results of on-the-fly generalization of line features using the Reactive-tree and the BLG-tree can be found in van Oosterom and Schenkelaars (1993): querying World Databank II (Gorny and Carter, 1987) (about 38 000 line features with total size 110 Mb) with response times below 5 s.

9.4.1 Example

In this section, an example of the use of the Reactive-tree, BLG-tree, and GAP-tree will be given. The Reactive-tree access method can be defined before any tuples are added to the relation, or one can first insert all tuples and decide later that a Reactive-tree is needed. The following two Postquel queries show the definition of the user table `AreaFeature` and the definition of a Reactive-tree index on this table using the `Reactive2_ops` operator class.

```
create AreaFeature (Height=int2, Idcode=int2,
Tree=int2,
  Roof=int2, shape=POLYGON2, reactive = REACTIVE2)\g

define index af_index on AreaFeature
  using ReactiveTree (reactive Reactive2_ops)\g
```

[2] DLMS (Digital Land Mass) DFAD (Digital Feature Analyses Data) Level 1 has a data density which can be compared to 1:200 000 scale map.

The GEO++ system automatically generates Postquel queries with the proper values for the BLG-tree and the Reactive-tree depending on the current scale. For the map in Figure 9.5, the following query is generated:

```
retrieve                (blg_pgn2=Blg2Pgn(AreaFeature.shape,
"0.01"::float4))
  where AreaFeature.reactive && "(13,40,23,47,2)"
::REACTIVE2
  sort by AreaFeature.oid
```

The Postgres query optimizer automatically selects the Reactive-tree access method when evaluating this query. The BLG-tree is used by specifying the function `Blg2Pgn` in the target list of the query. The linearized GAP-tree is reflected by the *sort by* clause of the query.

9.4.2 Benchmark

The visual results of the on-the-fly generalization techniques can be seen in Figures 9.5–9.7. All maps, including the overview in the upper right corner, are generated by the same query with only different-sized retrieved regions. DLMS DFAD can be regarded as land-use data with over 100 different area classifications, such as: lake, water, trees, sand, swamp, tundra, snow/ice, industry, commercial, recreational, residential, etc. A few notes with respect to the visualization follow.

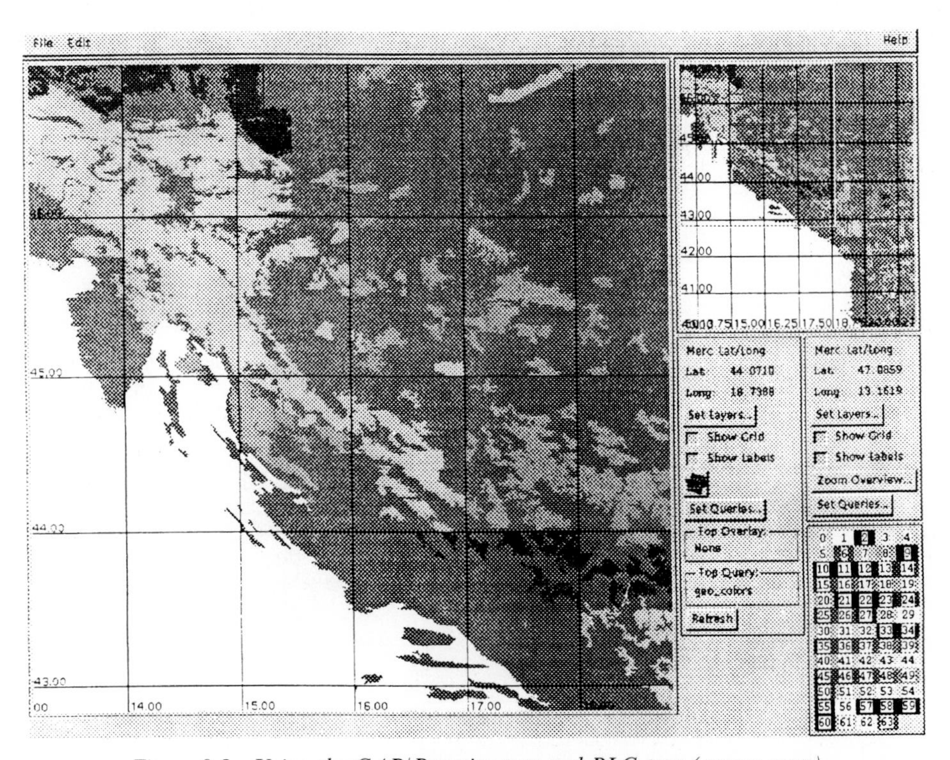

Figure 9.5 Using the GAP/Reactive-tree and BLG-tree (coarse map).

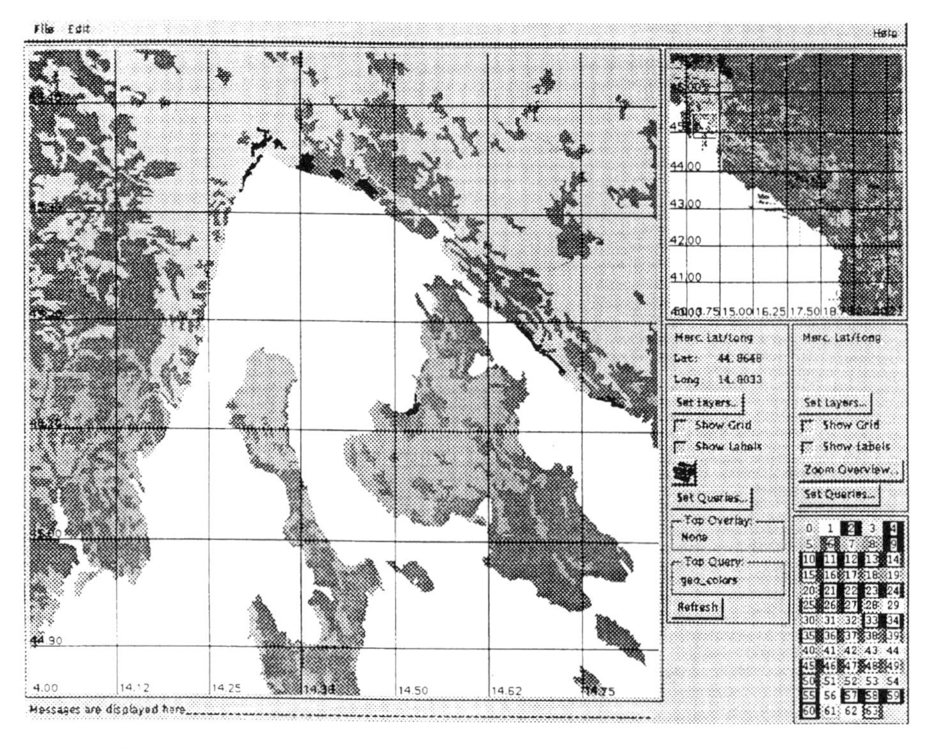

Figure 9.6 Using the GAP/Reactive-tree and BLG-tree (middle map).

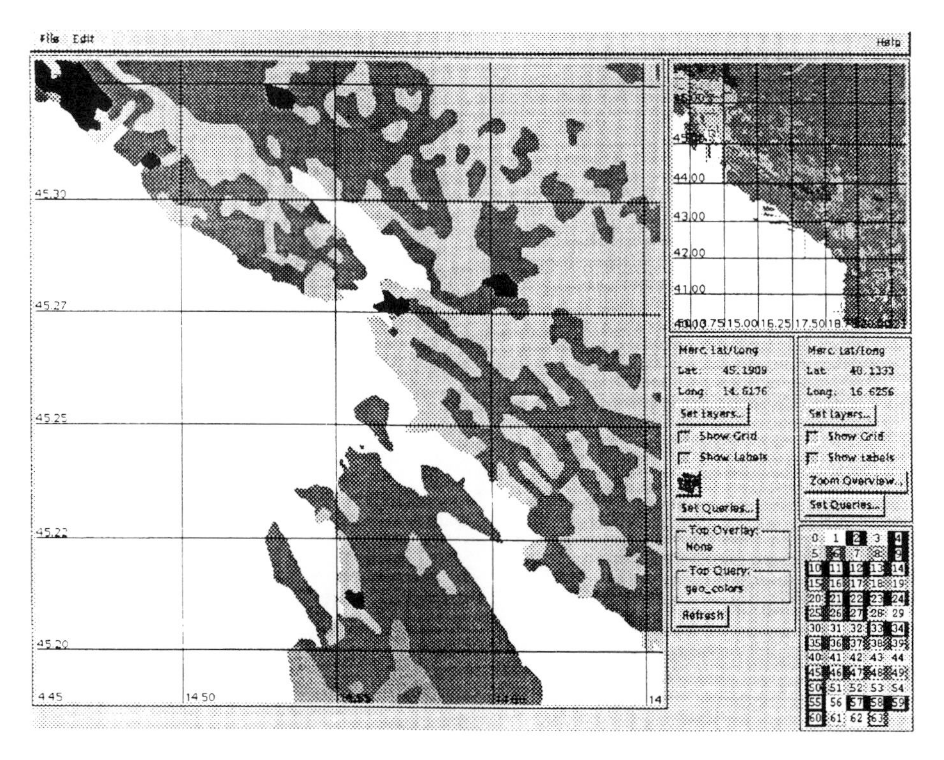

Figure 9.7 Using the GAP/Reactive tree and BLG-tree (detail map).

1. Many colours on the screen are lost in the grey scales of the printer.
2. As the emphasis is on the GAP-tree, the line and points features have been omitted, resulting in an incomplete map.
3. In the upper right corner of each figure, an overview of the region is shown without the mainland of Italy.
4. Though the DLMS DFAD data is stored in a seamless database, it has been digitized on a map sheet base. During this process, similar features have been classified differently; see Figure 9.5 near the 44° meridian.

The DLMS DFAD test database contains 70 272 area features (requiring about 60 Mb) which are given an importance value ranging from 1 to 5; at each level the more detailed level contains about one order of magnitude more features. Queries which selected tuples at five different area sizes were executed. The area sizes varied from $1.6 \times 1.6°$ to $0.1 \times 0.1°$ latitude/longitude. The Reactive-tree uses importance levels to reduce the number of selected objects at the more coarse level. Actually, the information density (i.e. the number of displayed features) should remain equal under the varying scales. The use of the GAP-tree will ensure that the map does not contain gaps when omitting the less important area features. The R-tree has no mechanism to select on importance level, which is why the R-tree retrieves many more objects in the larger areas. The following cases were tested: no index structure, an R-tree index, an R-tree index and a BLG-tree, and finally a GAP/Reactive-tree and a BLG-tree. In all cases, 10 random area queries were executed. The presented response time is in seconds and is the average time over the 10 random generated queries.

The data were retrieved over a network file system by Postgres. The special test program is a simple Postgres front-end application. The response time was measured with the Unix time command. The test program needs a parameter which indicates the size of the search area. This size is used to calculate which importance values are to be retrieved.

The difference in response time between the GAP/Reactive-tree and the R-tree decreases when moving to smaller areas, since the difference in the number of returned objects narrows. Table 9.1 shows the average number of returned objects for each method. In a similar way, the BLG-tree is more effective when used in a large area search. In that case, a lot of polygon points can be omitted.

All in all, the GAP/Reactive-tree and BLG-tree combination gives very good improvements. The process of selection and generalization results in a better map without too much detail. The selection could be done with the R-tree, but in that case the user should change the queries on every scale in order to select only the desired objects. Also, the important objects would then be located at the leaf level in the tree as were the other objects, which would slow down the small-scale queries. Figure 9.8 shows the response

Table 9.1 Area sizes and the average number of returned objects

Area (in square degrees)	0.01	0.04	0.16	0.64	2.56
No index	25.9	64.7	198.0	723.6	2262.3
R-tree	25.9	64.7	198.0	723.6	2262.3
R-tree and BLG-tree	25.9	64.7	198.0	723.6	2262.3
GAP/Reactive-tree and BLG-tree	25.9	52.8	43.6	55.8	48.7

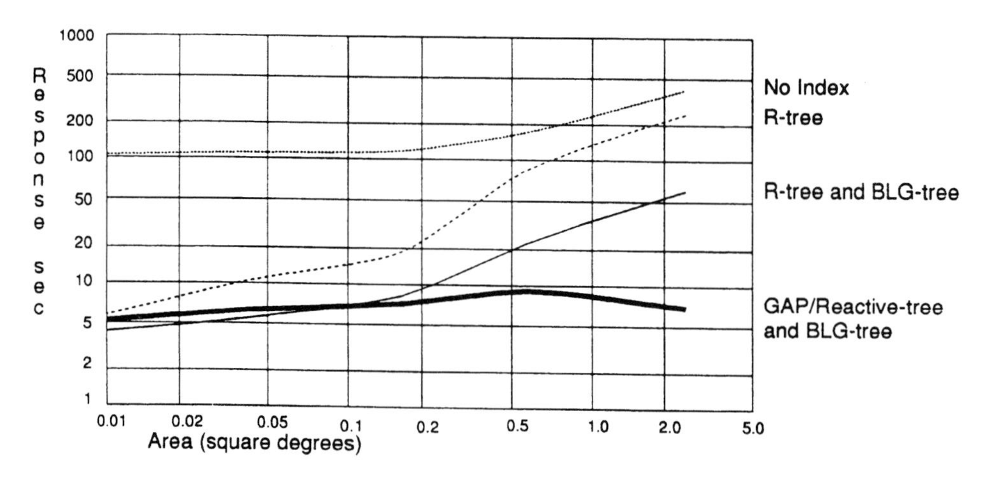

Figure 9.8 The average response time of the index structures.

Table 9.2 Area sizes and the average response time (including sort)

Area (in square degrees)	0.01	0.04	0.16	0.64	2.56
No index	113.8	117.7	127.1	183.4	378.7
R-tree	6.6	10.2	19.0	80.5	285.8
R-tree and BLG-tree	4.3	5.7	8.1	21.2	61.2
GAP/Reactive-tree and BLG-tree	5.4	6.9	7.5	9.7	7.3

times of the different methods, while Table 9.2 shows the same information in tabular form.

'If geometrically close objects are guaranteed to be stored close together on disk, some extra speed may be gained. The Reactive-tree does not give such guarantee. It depends on the inserting order whether two geometrically close objects are also stored close on disk. *Clustering* can be used to improve this storage aspect. In Postgres, a simple kind of clustering can be very easily simulated by retrieving the objects using a rectangle that contains all the objects. In fact, an access method scan is performed. When the objects are stored in a new relation in the retrieved order, geometrically close objects are stored close together on disk. This reduces the number of disk pages to be fetched for spatial queries and, therefore, the results will be returned faster.

9.5 Conclusions

After the Reactive-tree and BLG-tree, the GAP-tree forms a new important step towards the realization of an interactive multi-scale GIS. It is now possible to browse interactively through 'area partitioned' geographic datasets. In a database with more than 70 000 features it is now possible to get map displays at any required scale in about 6–9 s. In the near future, more tests with datasets are planned, e.g. with the Topographic base map of The Netherlands. An interesting open question is how to assign importance values automatically to features when building the Reactive-tree for a new dataset.

Further research is also required to determine how the GAP-tree can be maintained efficiently under edit operations. Additional further research topics are: the use of (dynamic) clustering techniques, and the design and implementation of other generalization techniques to support combination, symbolization, and displacement.

Acknowledgements

The author would like to thank the Postgres Research Group (University of California at Berkeley) for making their system available. Special thanks are also given to Tom Vijlbrief for helping with the GEO++ issues and to Vincent Schenkelaars for performing the DLMS benchmarks. Many valuable comments and suggestions on a preliminary version of this paper were made by Marcel van Hekken and Paul Strooper.

References

ATKIS, 1988, Arbeitsgemeinschaft der Vermessungsverwaltungen der Länder der Bundesrepublik Deutschland (AdV), *Amtliches Topographisch-Kartographisches Informationssystem (ATKIS)* (in German).

Ballard, D.H., 1981, Strip trees: a hierarchical representation for curves, *Communications of the ACM*, **24**(5), 310–21.

Boudriault, G., 1987, Topology in the TIGER file, in *Proceedings of Auto-Carto 8*, pp. 258–69.

DGIWG DIGEST, 1992, Digital geographic information — exchange standards — edition 1.1, Technical Report, Defense Mapping Agency, USA, Digital Geographic Information Working Group, October.

DMA, 1986, Product specifications for digital feature analysis data (DFAD): level 1 and level 2, Technical Report, Defense Mapping Agency, Aerospace Center, St Louis, MO.

Douglas, D.H. and Peucker, T.K., 1973, Algorithms for the reduction of points required to represent a digitized line or its caricature, *Canadian Cartographer*, **10**, 112–22.

Gorny, A.J. and Carter, R., 1987, World Data Bank II: General users guide, Technical Report, US Central Intelligence Agency, January.

Günther, O., 1988, *Efficient Structures for Geometric Data Management*, Number 337 in Lecture Notes in Computer Science, Berlin: Springer-Verlag.

Guttman, A., 1984, R-trees: a dynamic index structure for spatial searching, *ACM SIGMOD*, **13**, 47–57.

Jones, C.B. and Abraham, I.M., 1987, Line generalization in a global cartographic database, *Cartographica*, **24**(3), 32–45.

Lagrange, J.P., Ruas, A. and Bender, L., 1993, Survey on generalization, Technical Report, IGN, April.

Molenaar, M., 1989, Single valued vector maps: a concept in Geographic Information Systems, *Geo-Informationssysteme*, **2**(1), 18–26.

Peucker, T.K. and Chrisman, N., 1975, Cartographic data structures, *American Cartographer*, **2**(1), 55–69.

Projectgroep, 1982, IDECAP interactief pictorieel informatiesysteem voor demografische en planologische toepasingen: Een verkennend en vergelijkend onderzock. Technical Report Publicatiereeks 1982/2, Stichting Studiecentrum voor Vastgoedinformatie te Delft.

Robinson, A.H., Sale, R.D., Morrison, J.L. and Muehrcke, P.C., 1984, *Elements of Cartography*, 5th Edn, New York: John Wiley.

Shea, K.S. and McMaster, R.B., 1989, Cartographic generalization in a digital environment: when and how to generalize, in *Proceedings of Auto-Carto 9*, April, pp. 56–67.

Stonebraker, M., Rowe, L.A. and Hirohama, M., 1990, The implementation of Postgres, *IEEE Transactions on Knowledge and Data Engineering*, **2**(1), 125–42.
van den Bos, J., van Naelten, M. and Teunissen, W., 1984, IDECAP interactive pictorial information system for demographic and environmental planning applications, *Computer Graphics Forum*, **3**, 91–102.
van Oosterom, P., 1989, A reactive data structure for Geographic Information Systems, in *Proceedings of Auto-Carto 9*, April, pp. 665–74.
van Oosterom, P., 1991, The Reactive-tree: a storage structure for a seamless, scaleless geographic database, in *Proceedings of Auto-Carto 10*, March, pp. 393–407.
van Oosterom, P., 1993, *Reactive Data Structures for Geographic Information Systems*, Oxford: Oxford University Press.
van Oosterom, P. and Schenkelaars, V., 1993, Design and implementation of a multi-scale GIS, in *Proceedings EGIS'93: Fourth European Conference on Geographical Information Systems*, EGIS Foundation, March, pp. 712–22.
van Oosterom, P. and van den Bos, J., 1989, An object-oriented approach to the design of Geographic Information Systems, *Computers & Graphics*, **13**(4), 409–18.
Vijlbrief, T. and van Oosterom, P., 1992, The GEO system: an extensible GIS, in *Proceedings of the 5th International Symposium on Spatial Data Handling,* Charleston, SC, August. Columbus, OH: International Geographical Union IGU, pp. 40–50.

SECTION IV

Knowledge Acquisition

10

Potentials and limitations of artificial intelligence techniques applied to generalization

Stefan F. Keller

Unisys (Suisse), AG, Zuercherstrasse 59–61, CH-8800 Thalwil, Switzerland

10.1 Introduction

Over the past few years, the research interest in map generalization has been intensified so it is obvious that promising new techniques from other fields are being considered. While concentrating on issues of knowledge elicitation, problem solving and control, structure recognition, and techniques to implement cartographic operators, the author proposes to adapt further the research results of disciplines such as artificial intelligence (AI) and cognitive science including knowledge engineering (KE), work psychology (human–computer interaction) and social sciences.

Generalization will be defined as the process of abstraction used when the scale of a map is changed and the map content has to be visualized based on some communication purpose. This chapter will elaborate more on cartographic map generalization (in the sequence called simply 'generalization') than on model generalization, although the latter will be an interesting application for the utilities market too (e.g. in the production of geoschematic maps for electricity).

10.1.1 Knowledge-based approaches

In early attempts 'first generation' expert systems or knowledge-based systems (KBS) have not been a successful approach to the complex domain of map generalization. As in their original application in the AI community some years ago, there has been a disillusionment from the speculative promises. The common difficulties have now been identified and will be discussed. According to the basic building blocks of a KBS — knowledge representation, knowledge acquisition (KA) and implementation — all three components are affected.

The holistic approach to the generalization task can only be tackled considering the generalization problem as a joint human–machine task. Based on this idea, the term 'amplified intelligence' has been proposed (Weibel, 1991a). This approach can be extended by the so-called case-based reasoning (CBR) which supports, for example, the operator selection task by retrieving stored cases and respective solution plans to control the generalization process.

10.1.2 Knowledge in generalization

Past research in map generalization focused on rather narrow aspects of the complex problem of map generalization but, as stated above, recently there has been a trend towards more comprehensive and holistic solutions (Shea and McMaster, 1989, p. 56). To establish a holistic view there seems to be a common agreement that more structured knowledge has to be incorporated into the generalization process (Weibel, 1991b). It will be shown that although the need for more built-in knowledge seems to be obvious, this approach has some inherent limitations too. For example, KBS suffer from the 'decomposition problem' which denotes the dilemma of the structure recognition process that searches for patterns not knowing beforehand which are present. In the sections on neural nets (NN) and genetic algorithms (GA), these issues are explained.

KE methods currently receive high attention in map generalization research initiatives (Müller *et al.*, 1993; Weibel, 1993). Here, too, a side-view of similar studies in research on computer-aided design systems indicates already some strengths and weaknesses of proposed elicitation techniques which, in turn, could complement traditional KE studies. It has to be noted that up to now KBS have been built typically for analytical tasks like classification and diagnosis, rather than graphic design or construction.

10.1.3 The generalization task: operators, processes and evaluation

There are some fundamental problems with algorithmic procedural approaches. A cartographer always considers the complete relation of a feature with all other features in the surrounding area regardless of which 'layer' they are in. This relation is very difficult to establish for a system and there are not so many algorithms which put attention to these constraints. After having eventually solved the prior problem of relation, another arises when a decision has to be made: which feature should lead the generalization task if there are two adjacent linear or areal elements? This is the priority problem. For example, given a river which partially coincides with a state boundary, which one should be generalized first? Using amplified intelligence techniques, a convincing conceptual model has been laid out which enables knowledge engineers to gather structured data using, for example, transaction logs.

It will be shown that the development of cartographic operators and processes could profit from the application of new methods such as the neuro-biologically inspired NN, and the evolution-theoretic GA to line simplification, process control and feature evaluation. NN are showing stable behaviour in noisy domains and have been successfully applied to pattern matching. GA are proposed to be used for line simplification or line classification, as well as for the evaluation of existing algorithms by viewing these tasks as a suboptimization process.

10.1.4 How can generalization processes be evaluated?

The author supports the question of whether 'objective' generalization exists at all. Some publications make this assumption implicitly while emphasizing the diversity of maps (Spiess, 1990) while others begin to investigate this issue (Weibel, 1993, personal communication) and some are already convinced that this is the case (Mazur and Castner, 1990).

One pursued goal here is to provide the same solution as that of a human cartographer (Wang and Müller, 1993, p. 98). At least for training purposes of certain algorithms, this is an effective way to incorporate knowledge without the actual need to represent (and code) it explicitly. As an alternative, plans exist also to evaluate the results on existing algorithms as soon as 'objective' evaluation and measurement criteria are available. But a commonly accepted evaluation criterion for the performance of a generalized output remains to be established.

In the following sections, first successful and questionable promises of AI and KBS are discussed, then the advantages and pitfalls of KE methods with respect to knowledge elicitation are laid out so that the last section is centred on conceptual thoughts about the application of CBR, NN and GA. The discussion of the suitability and limitations of AI-related techniques is emphasized rather than the elaboration of technical details.

10.2 Knowledge systems

Ambitious research in AI produced the first commercially available KBS in the 80s. In order to extend the definition of KBS and to make it more amenable to neutral inspection the author prefers the term 'knowledge system'. Knowledge systems include the past notions of the same problem-solving technique (expert system, production system, knowledge based system, etc.) which can be defined by a specific architecture separating knowledge and inference and containing, for example, a representation scheme, a problem-solving control mechanism and an explanation facility. Additionally, this definition allows us to include systems like NN and CBR. Likewise — to stop the debate about what (human) intelligence is — there is an alternative to AI emerging called 'computational intelligence' with the following definition: on one hand AI is symbolic-type processing, or essentially high aspiration, duplicating or capturing some essence of human intelligence. On the other hand, computational intelligence includes techniques like NN, GA and other computational processes that exhibit the most abstract sort of intelligence in order to support human tasks, and probably tell us little about human intelligence itself.

Before more details are explained, some considerations of their respective limitations should be elaborated. In terms of knowledge systems, building a successful solution, in general, means first to define the representation of the generalization knowledge, then second to elicit the knowledge about operators and sequences, and finally to identify and control the processes.

10.2.1 Knowledge-based systems

The appealing idea of KBS is based on the logical and executable form of 'modus ponens' which looks like the common sense statements experts are using when describing, for example, 'IF the diameter of the reduced feature is lower than 0.2 mm THEN exaggerate it or delete it'. The straightforward implementation of knowledge as 'heuristic' rules extracted and coded this way led to the so-called 'first generation expert systems'. Despite some successful implementations in very narrow domains like medical diagnosis and technical configuration, all shared the following common shortcomings.

- In building KBS, KA became a critical phase which ended up in a KA bottleneck.
- KBS are not able to detect their own limitations and are therefore not able to explain the reason of the breakdown.

- With unexpected input, the inference process cannot deal with incomplete knowledge. This is a lack of adaptivity and of graceful degradation.
- KBS suffer from the decomposition problem.
- KBS are not capable of increasing performance and problem-solving competence with experience.
- The problems of structured representation are that KBS (and NN, too) have no powerful means of knowledge structuring. But this is a crucial part of building knowledge systems, so there is a lack of modularity and compositionality (Gutknecht, 1993, p. 81).
- Maintenance strategies have been neglected so that the smallest changes (like introducing a new rule) provoke the update of the whole rule base; so there are no comprehensive validation tools available.
- There is a shortage of knowledge engineers and sometimes ambiguous support from the management and traditional computer science establishment.
- There is reluctance from cartographic experts to contribute knowledge to prototype systems (Weibel, 1991a).
- There are restrictions to narrow domains which is the most important reason why we still wait for successful map generalization tools.

Several critiques from inside the AI community and from related fields like psychology, philosophy, cognitive and social sciences, helped to point to ways to overcome these drawbacks: philosophers observed two generic streams within AI in general, one that understood AI as a theoretical–philosophical discipline which uses the computer as a model of explanation and production of 'spirit' and 'problem understanding'. The others looked at AI as an application-oriented engineering discipline that wants to design automated supporting systems for knowledge-based information processing. For the purpose of map generalization the latter viewpoint seems to be the favourite choice but the former helps us to understand some common AI terms.

Connectionist researchers pointed to the lack of adaptivity, self-organization and learning. This results in an abrupt breakdown and has been called 'non-graceful degradation', i.e. the system will perform poorly with incomplete and noisy data. This has been partially solved with NN.

More recent critiques from the field of situated cognition — which stay in contrast to the rationalistic and tayloristic view — argue that KBS are not situated and dynamic in nature (Suchman, 1987). The relation of the system and the surrounding human world (the work situation) stresses the ongoing development process of this symbiosis instead of building more and more sophisticated systems. Exponents of work psychology also originated the term of 'tacit knowledge' (Johannessen, 1988). Therefore, 'every knowledge base in a sufficiently complex real world will always by necessity be incomplete' (Gutknecht, 1993, p. 126).

Cognitive engineers propose a more human-centred approach (Roth and Woods, 1989) using the so-called 'cognitive task analysis' and better human–computer interaction (Carroll and Olson, 1988; Gould and McGranaghan, 1990). Consequently, it would be more effective to automate such tasks at 100 per cent where computers are better than humans (like sequential search, calculations, storing, browsing and retrieving data) rather than pretending to automate the whole process but only at, say, 60 per cent success rate.

A solution of the incapabilities described by the situative and social critiques would be to include the user into the problem-solving cycle since he/she is present in the real-world situation and has (as an expert initially) a broader background of the domain. This interaction needs a thorough user-interface. It is, for example, crucial to be able to edit the whole rule-base and keep a good overview of it by using a query component for what-if analyses and a rule-base browser to retrieve related rules.

Newell (1980) was first to determine the 'symbolic level' of knowledge represented in KBS. It was also Newell (1982) who introduced the 'knowledge level' which, in turn, produced a 'second generation' of KBS. He stated that the widely used prototyping in KBS projects led to the mapping of human problem-solving competence via heuristic rules of thumb immediately to the symbol level where rules are coded. Instead, the knowledge level is a descriptive conceptual layer which investigates knowledge without taking care of rigid formalizations and operationalizations. With this goal in mind the knowledge engineer can concentrate on the character of the case and gets a better basis for reusable knowledge (Chandrasekaran, 1986). This approach leads to a neutral view regarding the implementation and results in a evaluation of different representations (rules, frames, semantic nets, tables, NN, GA, procedural code, etc.) and in structured knowledge.

Having analysed the underlying basic assumptions of KBS, knowledge systems and the conceptual model of amplified intelligence seem to be promising. This is a shift from batch solutions to interactive techniques which reflect the fundamental limitations of being not situated, lacking adaptivity and containing ill-structured knowledge. The lacking adaptivity and learning component is exactly what CBR wants to improve too, as will be explained below.

10.2.2 Knowledge representation in design tasks

At a first glance, the need for more built-in knowledge seems to be obvious. Several authors state, in a similar sense, that 'geographical generalization must incorporate information about the geometric structure of geographic phenomena' (Mark, 1989, p. 76), and 'the fact remains that procedural strategies that only consider the quantitative and geometric aspects of the generalization problem miss the point, because they cannot handle the information related to the semantic or attributes of geographical objects' (Mazur and Castner, 1990, p. 105). But it is not clear whether this knowledge should be exclusively incorporated in an explicit or implicit way. Partially because of perception, this top-down approach has to be combined with a bottom-up approach.

It is well known that perception plays a major role in generalization (Spiess, 1990). According to Shea and McMaster (1989, p. 57), six types of conditions for generalization exist. These conditions are highly subjective in nature and very difficult to quantify. In contrast to the currently proposed 'top-down' approaches, perception is rather a 'bottom-up' approach which needs to be handled by adequate algorithms like those found in low-level pattern matching.

Knowledge of a domain needs to be represented, but the same problem may have many representations. It seems, that 're-representation' of a problem can render it significantly easier to solve. If we know more about re-representation then we can choose the adequate representation for processing the related knowledge. The following questions arise therefore concerning, for example:

- the relation between alternative representations of the 'same' domain or problem;
- the mechanisms of concept formation and conceptual change involved in re-representation;
- the possibility of deriving a re-representation automatically by computer; and

- criteria for evaluating new forms of representation.

The key terms are 'representational redescription' (Clark, 1991) and 'knowledge compilation' (from Gutknecht, 1993, pp. 81–2).

Studies thought of as a criticism of connectionist systems (Clark, 1991) indicate that human learners are able to redescribe the compiled knowledge. In doing so, they decompile and explain it which is exactly what we require when looking for automated generalization architectures. The notion of 'cartographic compiling' is also mentioned to define what cartographers think of while resolving legibility problems (Müller *et al.*, 1993, p. 4).

Representational redescription is the inverse of what has been called the 'knowledge compilation' (Anderson, 1983).

> When acquiring skills, the learner first acquires explicit declarative knowledge (rules) about some tasks and interprets this knowledge when performing in the new domain. With more and more practice the knowledge gets transformed into a rapid procedural form. Instead of tediously interpreting pieces of declarative knowledge, the subject starts applying specialized procedures for entire subtasks. While this seems to describe the sequence, in which we are acquiring most of our knowledge at school (e.g. geometry), representational redescription seems to describe the sequence, in which we are acquiring knowledge and skills in behaviour oriented domains (e.g. drawing). In these domains we learn to perform and behave by imitation and trial and error (as in the medieval master–novice apprenticeship). Parallel to this there is a process of reflection and modular re-representation of the acquired knowledge . . . (Gutknecht, 1993, p. 82).

NN have difficulties in handling explicit representations (like knowledge compilation and explanation) but can learn very easily implicit knowledge from examples. In contrast, KBS have difficulties with learning but they can easily explain meta-rules (e.g. shortcut-rules) and make the knowledge base modular. Consequently, Gutknecht suggests the combination of both approaches, called 'Hybrid Systems', that can mutually compensate for their individual weaknesses.

An example of a hybrid system has been built by Hendler (1989), which combines a semantic network with a connectionist one. The NN 'grounds' the visual part to the leaf nodes of the semantic net, encoding perceptual features of these nodes. Those features he calls distributed 'microfeatural' representations of higher level concepts (from Gutknecht, 1993, p. 87). The architecture of 'Blackboard Systems' would be another approach to this problem (Craig, 1988).

10.2.3 Knowledge elicitation in design tasks

As stated above, KA methods currently receive high attention in research initiatives on map generalization (Müller *et al.*, 1993). Weibel (1993) gives a thorough analysis of the currently available generic KA techniques for cartographic generalization. This section focuses on elicitation of design knowledge in order to complement these techniques.

Knowledge elicitation in the domain of design, and particularly of visual and procedural knowledge, is very difficult (Tunnicliffe and Scrivener, 1991, p. 73; Gutknecht, 1993, p. 154). Two possibilities are mentioned by the comprehensive survey of KA methods by Welbank (1983, 1990): case study methods and protocol analysis.

There is also the clear notion of the borderline of spoken language which is called 'tacit knowledge' (Johannessen, 1988) and implicit knowledge (see, e.g. Welbank, 1983;

Dreyfus and Dreyfus, 1986; Tunnicliffe and Scrivener, 1991). In this sense, the communication of expertise is not simple either (Gaines, 1987). Because procedural knowledge seems to be rather tacit we have, for example, to look at the ongoing problem-solving process (e.g. the performance environment) or analyse recordings with visual and verbal protocols like video records of real cases.

The most direct and accurate source of information about knowledge is provided by the domain expert (Tunnicliffe and Scrivener, 1991, p. 83). This does not mean that we can bang rules out of the head of an expert. The transfer remains a difficult problem. This implies for protocol analysis that asking the right questions is very crucial for finding out the facets and quality of a given task. In contrast to the statement in Müller *et al.* (1993, p. 10), the fundamental issue in knowledge-based approaches is not 'whether we can represent generalization knowledge with "IF–THEN" production rules', but what the adequate combination of formalizations in generalization knowledge could be. This means that 'IF–THEN' rules are already a rough interpretation. Formalization is always combined with an interpretation process.

Gaines (1987) argues that pre-defining the knowledge representation and inferencing procedures before elicitation limits the usefulness and scope of knowledge system tools (the 'tool-trap' problem). Tunnicliffe and Scrivener (1991) argue also that the direct incorporation of data immediately into 'rules' is not appealing, 'especially in the absence of realistic information about the nature of design' (p. 74). So what is needed is a thorough evaluation of knowledge elicitation techniques for map generalization.

Already, many researchers are stating explicitly that there have been no sound elicitation procedures for design tasks developed so far (Tunnicliffe and Scrivener, 1991; Weibel, 1991b). In a study conducted by Tunnicliffe and Scrivener (1991), affirming the feasibility of knowledge elicitation, they advocate a method called the 'teachback method' based on the work of Johnson and Johnson (1987) which seems to be a successful approach. In the 'teachback method' the knowledge engineer and the domain expert have to sit together and communicate through verbal explanations and written sketches. The knowledge engineer has to replicate the case back until the expert is satisfied with the explanations. The resulting written documents are called 'mediating representations' because they reflect the intermediate and not pre-structured formalization of the knowledge.

It is worthwhile to mention a problem highlighted by research on situated cognition, which can occur if amplified intelligent systems are put into work environments of skilled or novice users: the notion of the 'changing target' problem (Suchman, 1987; Clancey, 1991). If map generalization is really viewed as a whole problem-solving process, the introduction of a new subsystem (our application) significantly changes the whole process itself, so that the target environment the system has been designed for is not the same any more. Now the operator has to work with a computer screen and a pointer device, he must make his hypotheses explicit, and he has to evaluate the provided suggestions and defaults.

The only solution of this problem is to take the environment into account when planning the KA process and 'viewing the knowledge system as a component of a joint user–computer system' (Gutknecht, 1993, p. 155). Those problems are tackled by an adaptive learning component.

10.3 Extensions to amplified intelligence

After the discussion of characteristic properties it is now worthwhile to look at the consequences. Regarding amplified intelligence from this point of view, this concept is not only an interactive approach but also a concept of cooperative systems giving back control to the user and being rather a decision support medium and a versatile tool. The ability of recording interaction logs promises more insight into the map generalization processes (as, for example, sequences of generalization operators and parameter estimation) given typical but realistic situations. On the other hand, this interactive approach does not exclude the fact that the system could propose a set of operators, for example (which have been pre-classified), or calculate adequate parameter defaults. This could be done by retrieving past cases or by a partial batch solution.

10.3.1 Case-based reasoning for generalization problem solving

Case-based reasoning (CBR) can be regarded as a subfield of machine learning. CBR is not only a particular reasoning method but also a machine learning paradigm that enables sustained learning by updating a case base after each completed problem–solution cycle. In CBR, learning occurs as a natural 'by-product'. When a problem is successfully solved, the 'case' is saved in order to solve it more efficiently next time. When the attempt to solve the problem fails, the reason for the failure is identified (if possible) and indexed in order to avoid the same mistake in the future. Effective case-based problem solving requires a sophisticated set of methods in order to extract relevant cases, integrate a case into an existing knowledge structure, and index the case for later matching with similar cases (Aamodt, 1991, p. 255).

It was shown that, particularly in early learning, people use past cases as schemes when solving unknown problems (Anderson, 1983). According to this theory, people generalize and operationalize (or compile) concrete experiences into chunks of production rules. Studies from less-structured domains (Kolodner, 1988) indicated that using past cases is a main problem-solving method among experts as well. While in the first case-based approaches a previous case was retrieved based on measurements of normalized distances among descriptors (Euclidean, similarity metric), more recent approaches try to match cases based on semantic similarities combined with inference strategies and indications on how strong the match is.

The high-level CBR cycle looks like this (see Figure 10.1):

1. Given a new case, find the best matching previous case (**remembering**).
2. Build a solution to the new problem by using and modifying the retrieved similar case (**transfer and adoption**).
3. Apply the modified solution plan and evaluate it (**testing and evaluation**).
4. Index and store the 'important' and eventually generalized information of this new solved case together with the problem-solving steps (**learning**).

In fact, there have been many different approaches to CBR which all implement the following methods in some degree: organizing, retrieving (matching), indexing and utilizing past cases. As a drawback, the case-based approach is sensitive to the number of attributes and has difficulties incorporating background knowledge.

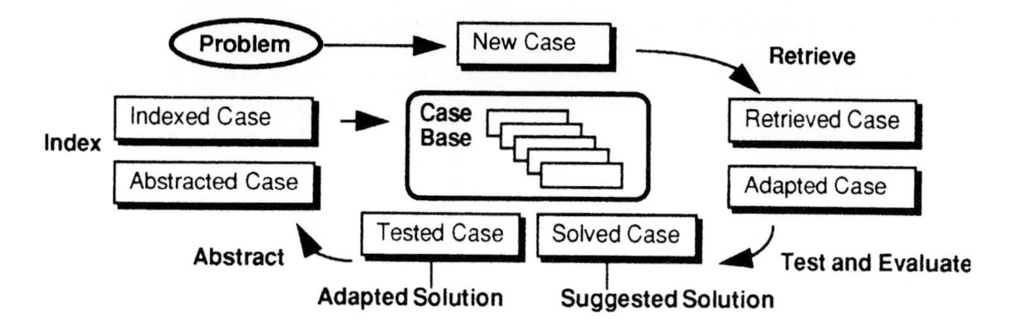

Figure 10.1 The CBR cycle (after Aamodt, 1991).

10.3.2 Neural nets for generalization representation

There have already been some first attempts to use NN in the GIS domain. The known strengths of NN are that the development time is shorter, they seem to be able to circumvent the KA process, and are stable with noisy input.

But NN are time intensive in supplying data, building the architecture, and have their limitations in high-level reasoning.

- **Small range of problem domains**: basically well suited for classification, pattern recognition, prediction, optimization, and control.
- **Poor explanation capabilities**: the explanation of a NN is based on the interpretation of primitive features like their dynamics, weights and energy landscapes. This is on a low level whereas a human user expects explanations at a higher level of the domain itself. The crucial decisions include the selection of training examples, the architecture and the learning algorithm.
- **Poor serial processing**: though sequential behaviour of NN has been demonstrated, NN are only suited for a limited complexity of sequential steps.
- **Weak possibilities to incorporate explicity already known domain knowledge.**
- **Real time problems** if learning cycle is included.

An application example: the operator 'line simplification'

The human cartographer has an imagination about the real-world representation of the objects and through his eyes he guesses the visual rules emphasizing important 'microfeatures'. There are no pre-defined control points known a priori, like with Nurbs. Template matching is hardly feasible, because there are too many 'real world templates'.

The idea is to train the net for the mapping function to translate a certain input vector pattern to the corresponding output vector. After the training phase, we could feed the net, for example, highway 999 at scale A from a different area to produce the corresponding generalized line at scale B. Another side-effect of this approach could be that the performance of the net could indicate which parameters are most successfully providing information to the mapping function. This can be used as an indicator to measures that contribute much information to detect 'important' points.

10.3.3 Genetic algorithms for generalization processes

Genetic algorithms (GA) are models of machine learning which are inspired by the metaphor of processes of evolution in nature. GA have their origins in robotics and optimization techniques and are viewed as a particular method from the broader field of 'evolutionary programming'. Common to all algorithms is the simulation of evolution of individual structures through processes of selection, mutation and reproduction. The processes are evaluated by the performance (or 'fitness') of individual structures in their collective environment (see, e.g. Holland, 1975; Goldberg, 1989).

GA apply an informed pseudo-parallel search in a parameter space. Search can be viewed both as a kind of problem solving and an optimizing process of given parameters. More precisely, GA use a general heuristic for exploration in a population of structures and come up relatively fast with a suboptimal solution. This near-optimum is close to an optimum with very high probability. GA are badly suited for problems that require explicit problem domain knowledge and where this knowledge is relatively easy to formalize. However, GA outperform other algorithms (e.g. linear programming) on problems which are NP-complete and where other problem-solving strategies failed. Although simplistic from a biologist's point of view, these algorithms provide robust and powerful adaptive search mechanisms (see Figure 10.2).

Given an example from the current domain, a particular structure scheme could be a set of parameters (represented through a character string) which define the Jenks algorithm (Jenks, 1979). As a result, we obtain a combination of the parameters which sufficiently fit a given cartographic line object. To summarize in an extendable list, GA could be used in generalization for:

- optimizing parameters of existing algorithms;
- calculating best defaults for given graphic features ; and
- classifying algorithms by applying quality measures to the output of an algorithm.

GA are constructive methods because they are algorithms that can be used to manipulate the configurational space of the design process, as opposed to the parametric space, to arrive at viable solutions. The manipulation and search of configuration space in engineering design has been considered as the creative aspect of design.

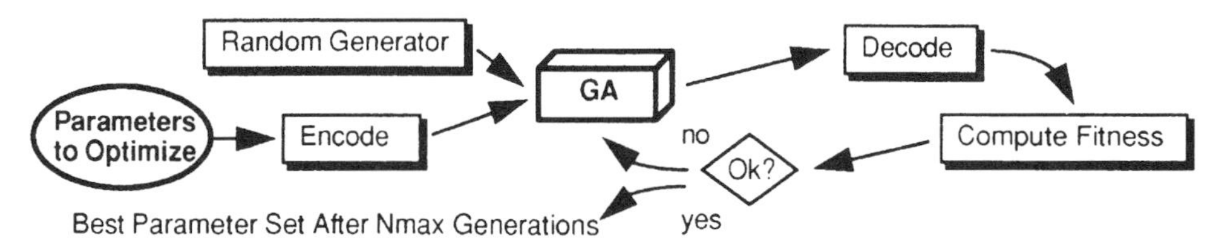

Figure 10.2 The data flow in GA.

10.3.4 Evaluation of NN and GA for cartographic generalization

Given the theoretically explained pitfalls of current AI techniques, where now are ways out of the current stagnation? A possible explanation is that AI techniques are not shaped yet for design tasks and the other is that they have been sometimes blindly adopted having in mind the complete automation of human tasks. The latter has been discussed above, the former will be elaborated upon now. So, what will be proposed is the combination of the two possibilities. The first is to continue investigation on existing methods but to refine them, and the second is to change the goal and to look for the intelligence from where it came, namely the human operator.

Therefore, generalization is being modelled on one hand as a bottom-up approach (e.g. Given a number of houses, what are the typical common features?) and on the other hand specialization is implemented choosing a top-down approach (e.g. Given the operator simplification, what is the specific instance of the relevant algorithm with the respective parameters?). For the whole generalization task, these two processes should be combined. The former is rather a low-level process which has to be tackled, for example, with NN. Preliminary experiments by the author, with limited resources, have shown results that can be visually interpreted.

NN adjust the representation of data structures through justification of internal parameters describing a mathematical function which maps input to output patterns and exhibits inter- and limited extrapolation. This is an inductive method called 'test-and-adapt', requiring many examples. GA adapt through the selection of operator sequences or the selection of data structures (algorithm parameters, measures). This is a constructive method 'propose-test-and-adapt'.

10.4 Conclusions

Many AI-related techniques offer great potential for the solution of generalization problems due to their adaptivity. Instead of pursuing mechanistic concepts such as the 'generalization machine' or 'objective generalization methods', the development should lead into the direction of cooperative knowledge systems which are based on a holistic view considering the generalization problem as a joint human–machine task.

Based on the examination of known properties, strengths and weaknesses of maturing AI techniques, two different approaches have been proposed. On one hand, the conceptual framework of amplified intelligence is extended by techniques of learning and managing past cases (CBR) combined with additional knowledge elicitation methods like the 'teachback method'. On the other hand, a sort of black-box approach is pursued based on manually generalized samples which aims to reveal inherent transformational functions in a re-representational form. Because of the complexity of the generalization task and the absence of adequate analytical methods, this dual approach can be described as a sort of constructive re-engineering, and preliminary feasibility studies have affirmed the basic theoretical expectations.

Keeping in mind the respective strengths, the following techniques have been presented as a promising research plan in map generalization: CBR for process control and problem-solving tasks, GA for complex non-predefined process sequences, and NN for open distributed and fuzzy representations. Therefore, this is a conceptual plan pursuing a combined strategy with interactive and batch systems at the same time using knowledge systems.

Acknowledgements

The author conducted this research while employed by Unisys (Suisse) S.A., which funded his travel. The opinions expressed in this paper are those of the author and do not necessarily represent those of Unisys (Suisse) S.A. The author is especially thankful for the discussions with Hansruedi Baer and the support from Robert Weibel and the author's family.

References

Aamodt, A., 1991, 'A knowledge-intensive, integrated approach to problem solving and sustained learning', unpublished PhD thesis, University of Trondheim, Norwegian Institute of Technology and SINTEF DELAB.

Anderson, J.R., 1983, *The Architecture of Cognition*, Cambridge, MA: Harvard University Press.

Carroll, M. and Olson, J.M., 1988, Mental models in human–computer interaction, in Helander, M. (Ed.) *Handbook of Human–Computer Interaction*, pp. 45–65, Amsterdam: Elsevier.

Chandrasekaran, B., 1986, Generic tasks in knowledge-based reasoning: high-level building blocks for expert systems design, *IEEE Expert*, **1**(3), 23–30.

Clancey, W.J., 1991, The frame of reference problem in cognitive modeling, in VanLehn, K. (Ed.) *Architectures for Intelligence. The 21st Carnegie Mellon Symposium on Cognition*, pp. 113–45, Hillsdale, NJ: Lawrence Erlbaum.

Clark, A., 1991, In defense of explicit rules, in Ramsey, W., Stich, P. and Rumelhart, D.E. (Eds) *Philosophy and Connectionist Theory*, pp. 115–28, Hillsdale, NJ: Lawrence Erlbaum.

Craig, I.D., 1988, Blackboard systems, *Artificial Intelligence Review*, **2**, 103–8.

Dreyfus, H.L. and Dreyfus, S.E., 1986, *Mind over Machine*, New York: The Free Press.

Gaines, B.R., 1987, An overview of knowledge acquisition and transfer, *International Journal on Man–Machine Studies*, **26**(4).

Goldberg, D.E., 1989, *Genetic Algorithms in Search, Optimization and Machine Learning*, Reading: Addison-Wesley.

Gould, M.D. and McGranaghan, M., 1990, Metaphor in geographic information systems, in *Proceedings of the 4th International Symposium on Spatial Data Handling*, Zurich, Vol. 1, pp. 433–42.

Gutknecht, M., 1993, 'Adaptive hybrid artifacts: three perspectives on designing artificial systems', PhD thesis, Institute for Informatics, University of Zurich.

Hendler, J.A., 1989, Marker passing over microfeatures: towards a hybrid symbolic/connectionist model, *Cognitive Science*, **13**, 79–106.

Holland, J.H., 1975, *Adaption in Natural and Artificial Systems*, Ann Arbor, MI: University of Michigan Press.

Jenks, G.F., 1979, Thoughts on line generalization, in *Proceedings of the International Symposium on Cartography and Computing, AutoCarto IV*, Vol. 4, pp. 209–20.

Johannessen, K.S., 1988, Rule following and tacit knowledge, *AI & Society*, **2**, 341–50.

Johnson, L. and Johnson, N.E., 1987, Knowledge elicitation involving teachback interviewing, in Kidd, A. (Ed.) *Knowledge Elicitation for Expert Systems*, New York: Plenum Press.

Kolodner, J. (Ed.), 1988, Case-Based Reasoning, *Proceedings from a Workshop*, Clearwater Beach, Florida, May, Los Altos: Morgan Kaufmann.

Mark, D.M., 1989, Conceptual basis for geographic line generalization, in *Proceedings of the Autocarto 9, 9th International Sympsium on Computer Assisted Cartography*, Baltimore, USA, April, pp. 68–77.

Mazur, E.R. and Castner, H.W., 1990, Horton's ordering scheme and the generalizing of river networks, *The Cartographic Journal*, **27**(2), 104–12.

Müller, J.C., Weibel, R., Lagrange, J.P. and Salgé, F., 1993, Generalization: state of the art and issues. Technical Report and Position Paper, *European Science Foundation, GISDATA Task Force on Generalization*, Compiegne, France.

Newell, A., 1980, Physical symbol systems, *Cognitive Science*, **4**, 135–83.

Newell, A., 1982, The knowledge level, *Artifical Intelligence*, **18**, 87–127.

Roth, E.M. and Woods, D.E., 1989, Cognitive task analysis: an approach to knowledge acquisition for intelligent system design, in Guida, G. and Tasso, C. (Eds) *Topics in Expert System Design*, pp. 233–64, Amsterdam: North-Holland.

Shea, K.S. and McMaster, R.B., 1989, Cartographic generalization in a digital environment: when and how to generalize, in *Proceedings of the Autocarto 9, 9th International Symposium on Computer Assisted Cartography*, Baltimore, USA, April, pp. 56–67.

Spiess, E., 1990, Bemerkungen zu Wissensbasierten Systemen für die Kartographie, *Vermessung, Photogrammetrie und Kulturtechnik*, **2**, 75–81.

Suchman, L.A., 1987, *Plans and Situated Actions*, Cambridge, MA: Cambridge University Press.

Tunnicliffe, A.J. and Scrivener, S.A.R., 1991, Knowledge elicitation in design, *Design Studies*, **12**(2), 73–80.

Wang, Z. and Müller, J.C., 1993, Complex coastline generalization, *Cartography and Geographic Information Systems*, **20**(2), 96–106.

Weibel, R., 1991a, Amplified intelligence and rule-based systems, in Buttenfield, B.P. and McMaster, R.B. (Eds) *Map Generalization: Making Rules for Knowledge Representation*, pp. 172–86, London: Longman.

Weibel, R., 1991b, Specification for a platform to support research in map generalization, in Rybaczuk, K. and Blakemoore, M. (Eds) *Proceedings of the 15th Conference of the International Cartographic Association*, Bournemouth.

Weibel, R., 1993, 'Knowledge acquisition for map generalization: methods and prospects', Position Paper in *Proccedings of Specialist Meeting for Initiative 8 'Formalizing Cartographic Knowledge'*, National Center for Geographic Information and Analysis (NCGIA), Buffalo, NY, 23–27 October, pp. 223–32.

Welbank, A.M., 1983, *British Telecom Report on Knowledge Acquisition*, Technical Report, British Telecom, London.

Welbank, A.M., 1990, An overview of knowledge acquisition methods, *Interacting with Computers*, **2**(1), 83–91.

11

Rule-orientated definition of the small area 'selection' and 'combination' steps of the generalization procedure

Michael Heisser, Georg Vickus and Johannes Schoppmeyer

Institut für Kartographie und Topographie, Universität Bonn, Meckenheimer Allee 172, 53115 Bonn, Germany

11.1 Introduction

ATKIS, the official Authoritative Topographical Cartographic Information System (Amtliches Topographisch–Kartographisches Informationssystem), is currently being created in Germany. ATKIS is made up of Digital Landscape Models (DLM) and Digital Cartographic Models (DKM) (AdV, 1989). The DLMs contain reproductions of topographical features and the relief of the earth's surface. DLMs are created as DLM 25 (based on a scale of 1:25 000), DLM 200 (= 1:200 000) and DLM 1000 (= 1:1 million), depending on the various requirements regarding semantic and geometric precision. The data structure is based on the individual topographic object. A topographic object is considered here to be the digital image of a topographic feature of the real world. In the following, the term 'topographic object' is substituted by the term 'object'. The cartographic generalized and variously symbolized map features derived from the individual DLM features then form the DKM.

Generalization of DKMs derived for various scales has to be undertaken in two domains (cf. Figure 11.1).

1. Feature generalization according to a symbol catalogue to create a DKM for the first time. This 'model generalization' transforms the DLM objects to the data structure of the DKM.
2. Cartographic generalization for the elimination of graphical presentation problems and to increase the legibility as a part of cartographic compilation, starting on the raw DKM via Digital Intermediate Models (DZM) and finishing on printed maps (Beines, 1993).

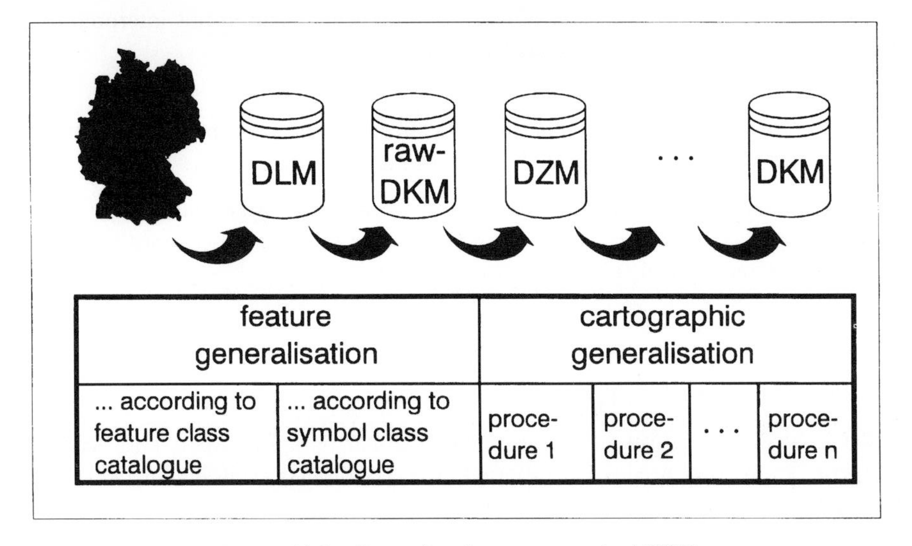

Figure 11.1 Generalization processes in ATKIS.

11.2 Cartographic generalization and knowledge-based techniques

Cartographic generalization is currently the main problem being encountered during the derivation of digital cartographic models. A fundamental concept for this process in ATKIS was developed in a research and development project carried out by the North Rhine–Westphalian State Surveyor's Office and the Institute for Cartography and Topography at the University of Bonn (Vickus, 1993). The concept is based on a procedure-orientated approach and contains the following individual tasks:

- definition of the scope of processing,
- structuring of the processing steps,
- algorithm development,
- definition of a processing order, and
- hierarchization and classification of the map feature classes in respect of the processing steps.

Although a procedural system is capable of generalizing a large variety of standard situations, in the long term, interactive revision work will always be required to solve special cases. The scope of this revision work should be kept to as bare a minimum as possible.

11.2.1 Problem definition/theses

The generalization of a map feature is dependent upon its significance, its position in relation to adjacent map features and its graphical presentation. The problem in cartographic generalization consists in the fact that the multitude of possible feature constellations on the map can be hardly registered by a purely procedure-orientated system. Moreover, the significance of a feature is difficult to model in some cases. The correct decision in cartographic generalization is consequently dependent upon the interpretation of the specific situation. In the case of conventional generalization this interpretation is based on the cartographer's knowledge and expertise.

The main question is therefore:

> What knowledge the cartographer applies during generalization and is a better perception and consideration of that knowledge possible?

The cartographer's knowledge and expertise has to be transformed in rules and procedures and made available to the cartographic generalization process with a process library (Brassel and Weibel, 1988; McMaster, 1991).

11.2.2 Types of knowledge applied in generalization

In order to evaluate the formalization potential of cartographic expert knowledge, the global term 'knowledge' needs to be differentiated. To this end, the following will attempt to separate and describe the individual types of knowledge.

1. Knowledge of geometry and graphics
   (a) geometry, coordinates and links
   (b) graphical design of features
2. Semantic knowledge
   (a) significance of individual feature classes
   (b) significance of individual features
   (c) significance of adjacent relations
3. Procedural knowledge
   (a) number of processing steps
   (b) processing hierarchy
   (c) algorithms
4. Structural knowledge
   (a) data structures
   (b) structure of individual features (e.g. meanders)
   (c) interaction between individual features on the map
      — density
      — topological relations
      — geometrical relations
5. Interdisciplinary expert knowledge
   (a) expert knowledge derived from the fields of geomorphology, hydrology, human geography, etc.

This structure leads to the hypothesis:

> All types of knowledge contained in this structure need to be represented in a coherent framework established for the creation of a 'knowledge-based' generalization system.

In order to evaluate the chances of such a concept succeeding, it is first necessary to answer the following questions:

1. What knowledge is already available?
2. How can knowledge not yet available be acquired?

1. The knowledge on geometric feature modelling and the graphical design of features has been amply defined in the ATKIS project by the feature class catalogue and the symbol catalogue. Even if both catalogues are updated in accordance with newly acquired knowledge, they nevertheless remain suitable for use as a working hypothesis.

In the end, procedural knowledge applies to the processing steps, algorithms and hierarchies that are used in a procedural system and without which a knowledge-based system cannot work.

2. The greatest demand on the knowledge and expertise of generalizing cartographers arises when it comes to inferring or reasoning. This is the case, above all, because problem solving is highly context dependent and not all constellations on the map can be traced back to standard cases. This means that good generalization requires exceptional professional experience and great graphical skill. It cannot be expected that the expert's inference knowledge and strategic knowledge can be completely analysed through job observation and interview. The methodological approach must take this difficulty into account. The cartographer's problem-solving ability can only be derived from the results of his or her work.

The available knowledge and expertise needs to be defined in a conceptional model. In order to keep the various types of knowledge generally valid, the available presentation formalisms should initially not be restrictive.

The recording of the generalizing cartographer's knowledge and expertise represents an important cornerstone for the use of knowledge-based software methods. The decision in favour of special representation formalisms must be appropriate to the acquisition of knowledge.

11.3 Rule-orientated definition of small area 'selection' and 'combination'

11.3.1 Starting point

Definition should be generally valid and system independent — therefore the use of special formalisms for knowledge representation will be dispensed with. More detailed explanation of this will be given using the example of the generalization steps of the areas 'selection' and 'combination'.

The information requiring further specification in this section can be classified into three categories.

1. Declarative model
   (a) Problem formulation
   (b) Data structures
   (c) Initialization information
2. Functional model
   (a) Characteristics and methods for data elements
   (b) System orders
3. Rule-orientated model
   (a) Rules
   (b) Inference mechanisms

11.3.2 Declarative model

Problem definition

For some feature classes, minimum area sizes have been prescribed for their transfer into the DKM. If a map feature does not fulfil such minimum area size requirements (e.g. 3 ha), then this map feature (cf. Figure 11.2: grey area R) has to be combined with an adjacent map feature (M or I, because these areas are not separated by a linear feature class). Its digital image, the object part, becomes a class 8 areal object part. That means this object part requires processing in cartographic generalization.

Various decision criteria exist for carrying out 'combination'. These will be explained in the following section.

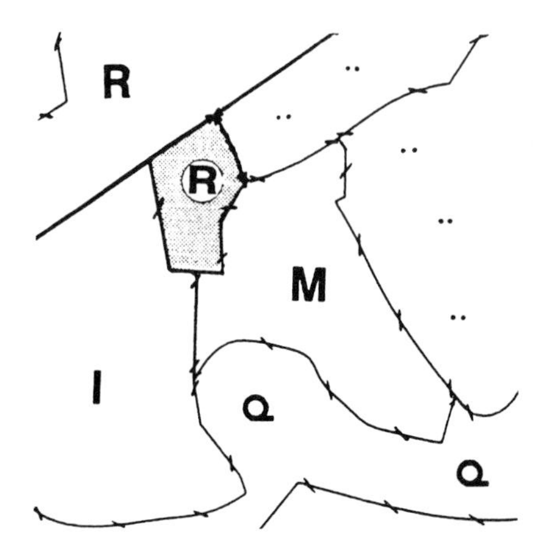

Legend (map feature class)

- **R** residential area (201)
- **I** general and light industrial area (202)
- **M** area of mixed use (203)
- •• meadow land (402)
- **Q** leafy wooded area (409)
- ↔ road (301)
- —/— border line without linear map feature class

Figure 11.2 Small residential area (grey) requires processing (extract from DGK 5 'Mühlheim a. d. Ruhr' — plot of linear geometry of DLM 25).

Feature classification and abbreviations

The coding system of ATKIS modelling is structured into feature classes, feature groups and feature categories. For each DLM feature class (OA), which will be represented in the DKM, there exists an appropriate DKM map feature class (KOBA), defined in the symbol class catalogue.

Within the scope of the rule descriptions, the following abbreviations, according to ATKIS standards, are used:

OBJ	ATKIS-Object
OT	ATKIS-Object part
VE	ATKIS-Vector element
KOBA	Map feature class
OA	Feature class in accordance with feature class catalogue
OT8	Class 8 areal OT (OT requiring processing)
OTN	Topologically adjacent areal OT
AZK	Selection and combination criteria (that means minimum area size in derivation process)

Data structure for areal features

Areal features are described by the following elements of the ATKIS data model:

- *Object* forms the semantic unit of one or more areal object parts
  ⇓
- *Object parts* are the essential unit during 'combination'. Object parts which fall below the minimum area size criterion during the derivation process from DLM to DKM are nevertheless depicted, but are marked by a flag in the object part class (Object part class 8)
  ⇓
- *Vector elements* are the area border lines for areal object parts, that line geometries

Initialization information

In order to 'select' and 'combine', some data need to be made available during initialization of the generalization process.

1. The assignment hierarchy defines to which adjacent object parts (OTN) the object part will be assigned if several of these OTNs belong to the same feature group as the object part undergoing processing.

*Assignment list (1:*n*, extract of feature group 'built-up area')*

Map feature class		Assignment hierarchy
201 (Residential area)	→	203, 202, 204
202 (General and light industrial area)	→	203, 204, 201
203 (Area of mixed use)	→	204, 202, 201
204 (Area of special use)	→	203, 202, 201

2. For feature categories 2000 (Settlements) and 4000 (Vegetation), the feature category assignment determines which feature category has the highest priority for 'combination' if no adjacent object part from the feature category comes from the object part undergoing processing.

Feature category assignment

Feature category		Feature category
2000 (Settlements)	→	4000
4000 (Vegetation)	→	2000
5000 (Hydrography)	→	—

3. The area overlying list determines which other areas have been defined beneath a given area so that the given area can simply be omitted in the course of selection. However, direct referencing in the data model would be easier.

Overling area (1:1), extract of feature group 'bulit-up area')

Map feature class	Overlies other area
201 (Residential area)	no
202 (General and light industrial area)	no
...	
213 (Large sewage plant)	yes

11.3.3 Functional model

Attributes and methods on object parts

The functional approach describes the methods with which the required interrelations and interactions resulting from of the constellation of the map features can be calculated. Some of the methods described here will be needed for rule-orientated problem-solving strategies.

- Number of object parts in the primary object (ANZ): the number of object parts belonging to the current object are counted.

$$\text{ANZ} = f\,[\text{Obj (OT8)}]$$

  The number of object parts is needed in order to determine whether combination of the object part undergoing processing (OT8) is possible with an object part of its own object.
- Combination priority (ZFP): the combination priority expresses how suitable for 'combination' the adjacent area is on the basis of its characteristics. This depends on the symbol section number (STN) of the primary object part (OT8) and the symbol section number of the adjacent object part, and is assigned to this.

$$\text{ZFP} = f\,[\text{STN(OT8), STN(OTN)}]$$

  The combination priority is:

  100 for the same map feature class,
  90 for the same feature group,
  80 for the same feature category,
  10 all other cases,

  and can be determined from the coding of the DLM feature class.
- Direct neighbours (DN): those adjacent object parts are determined whose common border line is not simultaneously a separating linear transport route or waterway, because combination is not permitted over a road or a waterway.

$$\text{DN} = f\,\{\text{OA [VE (OT8, OTN)]}\}$$

  The result is that either none, one or several object parts can be combined.

Moreover, other criteria also apply, such as the longest edge criterion, area overlying criterion, semantic criteria governing assignment, and the selection and combination criteria of model definition.

System orders

The global system order in this generalization step is to process all object parts with class 8 object parts to such an extent that the generalization steps 'selection' and 'combination' have been completed for these object parts and class identification (8) can be deleted.

Depending on the specific case, the global order can be carried out by means of three different special system orders.

1. OMIT: delete OT8 without replacement;
2. RETAIN: delete class identification without replacement; and
3. COMBINE: with an adjacent object part.

Whilst orders 1 and 2 complete the processing of the object part, operators must carry out a check after completion of order 3 to confirm that the combined object part has fulfilled the minimum area criterion for 'selection' and 'combination'. If this is not the case, then the combined new object part must again be subjected to the processing procedure as an OT8 (Iteration).

11.3.4 Rule-orientated model

Rules applicable to 'combination'

In the following, the individual decision-making criteria for 'combination' will be expressed as IF–THEN rules.

Rule to differentiate between the special orders (1)[1]

```
IF     area OT8 has no direct neighbours,
AND    another area is overlaid,
THEN   OMIT

IF     area OT8 has no direct neighbours,
AND    no other area is overlaid,
THEN   RETAIN

IF     area OT8 has some direct neighbours,
THEN   COMBINE
```

Whilst OMIT and RETAIN can be executed directly, a suitable adjacent object part must first be sought for COMBINE. The necessary reduction of all adjacent object parts onto a single object part is expressed in the following rule (2).

```
IF     only one object part is (still) 'suitable',
THEN   COMBINE
```

The following rules can be used for the selection and reduction of suitable object parts. The list does not represent any particular processing order, but is rather a list of necessary rules which can be changed at any time. The linking of individual rules only occurs as a result of the problem-solving strategy (cf. *Problem-solving strategy*).

[1] Numbers in parentheses refer to Figure 11.3.

Rules for reducing possible adjacent object parts

(a) Reduction according to the internal structure of the object (3).

```
IF    the object of the OT8 contains several OTs,
THEN  only these are suitable.

IF    the object of the OT8 contains only one OT,
THEN  all neighbouring OTNs are suitable.
```

(b) Topological reduction (4).

```
IF    an OTN is a direct neighbour of OT8,
THEN  it is suitable for 'combination'.
```

(c) Semantic–geometrical differentiation (5).

```
IF    several OTN object parts are suitable,
THEN  determine one or several with the highest ZFP.

IF    several OTNs simultaneously have the highest ZFP of 100,
THEN  select an OTN according to LKK.

IF    several OTNs simultaneously have the highest ZFP of 90,
THEN  sort according to ZL.

IF    several OTNs are suitable according to ZL,
THEN  select an OTN according to LKK.

IF    several OTNs simultaneously have the highest ZFP of 80,
THEN  select an OTN according to LKK.

IF    several OTNs simultaneously have the highest ZFP of 10,
THEN  select one (or several) OTNs according to LKK.
```

(d) Geometrical differentiation (default value) (6).

```
IF    several LKKs are identical[2]
THEN  select the first.
```

(e) Result evaluation (7).

```
IF    the (new) object part meets the AZK criteria,
THEN  finish processing of this object part, that is delete class
      identification 8.

IF    the (new) object part does not meet the AZK criteria,
THEN  process the object part again (→ 1).
```

[2] Geometric differentiation only covers those rare cases in which several edges are the same length, meaning that more than one OTN seems suitable according to the LKK criterion.

As an alternative to this, another approach will be considered in which the semantic and geometric characteristics are considered jointly, albeit weighted separately. This corresponds to Fuzzy-Set-Theory procedure.

In a combined approach (Fuzzy-Logic), weights (empirically determined) are introduced for the valency of the above-mentioned assignment stages and are assigned to all possible adjacent object parts. Additionally, further criteria can be called upon with which the adjacent object part with the highest valency, that is with probably the best degree of combination suitability, can be determined. For this purpose, a matching value (standardized to the value range of zero to 1) will be determined for each possible adjacent object part for each criterion. Individual criteria could be, for example:

- Longest edge criterion (LKK): the individual edges are placed in relation to each other and relative to the longest edge. The result is a value between zero and 1.

$$\text{LKK} = \frac{\text{Edge length}_i}{\text{Longest edge length}}$$

- Compactness (KOM): the ratio area:circumference (related to a standardized area) should be as large as possible. In this case, too, the individual ratios should be set in relationship to the largest ratio in such a way that the result is a value between zero and 1.

$$\text{KOM} = (\text{Area } i/\text{Circumference } i) \times \text{standardization factor}$$

- Assignment lists (ZL): the value for the quality of an assignment can be derived from the distance in the assignment list (cf. *Problem-solving strategy*) which the map feature class of an OTN has from the map feature class of the OT8. The values have been empirically specified.

$$\begin{aligned} \text{Distance 1: ZL} &= 0.8 \\ \text{Distance 2: ZL} &= 0.5 \\ \text{Distance 3: ZL} &= 0.0 \end{aligned}$$

Using these three criteria it is possible to determine a weighted overall matching value (ZGW).

$$\text{ZGW} = [(\text{LKK}/2) + \text{KOM} + \text{ZL}]/2.5$$

This approach is based on Fuzzy-Set-Theory and additionally enables easy integration of further criteria plus easy amendment of the generalization behaviour by amendment of the weighting.

Problem-solving strategy

'Combination' requires that the empty area assignment[3] has been completed, meaning that a map feature class is present for the whole map area.

---

[3] Empty area assignment guarantees that each area has a semantic definition (map feature class) independent of area size.

The problem-solving strategy for generalization is, in fact, the rational linking of individual rules, that is an inference mechanism. A suitable inference mechanism can be seen in Figure 11.3. It should be noted that this inference mechanism is an abstract depiction and does not refer to any implementation. The way in which the individual rules are applied, and which actions they involve, remains a matter for the respective realization. The numbers shown in parentheses in Figure 11.3 refer to matching sets of rules. The individual system orders are stressed by their thick borders. On the basis of superior rules (1), a decision is made as to which system order is to be carried out.

The purpose of 'combination' is to reduce constantly the number of 'suitable' adjacent object parts until just one is left (2). 'Combination' represents hierarchical selection, in which primary selection is carried out on the basis of semantic criteria and secondary selection on the basis of geometric criteria. When Fuzzy-Sets are applied, these criteria will be standardized and placed next to each other on an equal footing for selection. The essential difference between the simple and the mixed approach can be found on the semantic–geometric level (5). The former approach to solving this assumes strictly hierarchical questioning of individual criteria. If one of the criteria results in a decision, the subordinate criteria are no longer taken into consideration. In contrast to this, the mixed approach assumes that there are only two decision-making levels. On the first decision-making level, an attempt is made to reach a final and clear-cut decision by assigning combination priorities (ZFPs). Map sheet examinations have shown that even if only the combination priorities are assigned, this results in clear-cut and rational solutions in most cases. The separation of this first decision-making level would therefore seem plausible for the sake of simple processing. On the second level, all other decision criteria are taken into account. They are qualified and jointly considered.

For example, the LKK criterion has been formerly viewed as an absolute. The result was *the* object part with the longest common edge. If one takes a quantitative look, each edge (also meaning each adjacent object part) is assigned a relative length value. Similarly long edges therefore receive similar definitions.

Using the overall matching value it is possible to determine the adjacent object part which is best suited for combination when all three criteria have been taken into consideration.

11.4 Outlook

The rules described in this chapter are a contribution towards compiling a set of rules for cartographic generalization. Not only do the individual steps need to be modelled for the development of a complex knowledge-based generalization system, but so do, in particular, the interactions between these steps.

References

AdV, 1989, Amtliches Topographisch–Kartographisches Informationssystem (ATKIS), Das Vorhaben der Landesvermessungsverwaltung zum Aufbau Digitaler Landschaftsmodelle und Digitaler Kartenmodelle, Bonn: Arbeitsgemeinschaft der Vermessungsverwaltungen der Bundesrepublik Deutschland (AdV).

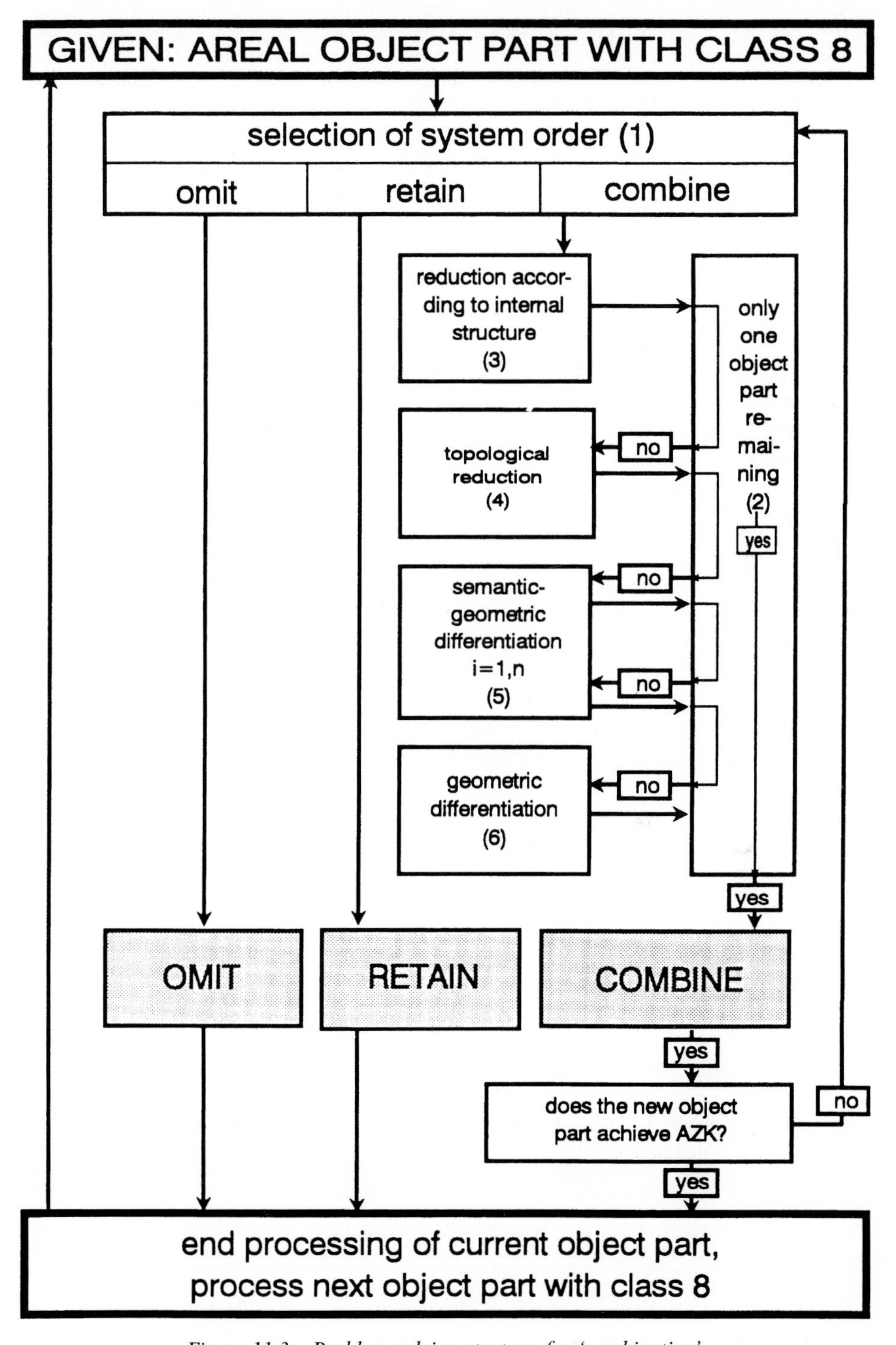

Figure 11.3 Problem-solving strategy for 'combination'.

Beines, M., 1993, Treating of area features concerning the derivation of digital cartographic models, in *Proceedings of the 16th International Cartographic Conference*, Köln, pp. 372–82.

Brassel, K.E. and Weibel, R., 1988, A review and conceptual framework of automated map generalisation, *International Journal of Geographical Information Systems*, **2**, 229–44.

McMaster, R.B., 1991, Conceptual framework for geographical knowledge, in Buttenfield, B.P. and McMaster, R.B. (Eds), *Map Generalization: Making Rules for Knowledge Representation*, pp. 21–39, London: Longman.

Vickus, G., 1993, Digitale topographische und kartographische Modelle sowie Entwicklung ihrer Überführungsstrukturen am Beispiel von ATKIS, Dissertation — Schriftenreihe des Instituts für Kartographie und Topographie der Universität Bonn, Heft 20, Bonn.

12

Knowledge acquisition for cartographic generalization: experimental methods

Robert B. McMaster

Department of Geography, University of Minnesota, 414 Social Sciences Building, Minneapolis, Minnesota 55455, USA

12.1 Introduction

Although research efforts in the area of digital cartographic generalization made excellent progress during the period 1965–80, work has slowed over the past 5 years. Most of the initial progress resulted from activity in the development of algorithms (such as the well-known Douglas–Peucker line simplification routine, a variety of techniques for smoothing data, and algorithms for displacement), as well as attempts to analyze both the geometric and perceptual quality of those algorithms. Recent attempts at the development of a more comprehensive approach for generalizing map features — such as the application of simplification, smoothing, and enhancement routines either iteratively or simultaneously — have not, for the most part, been successful. This is a result, in part, of a lack of procedural information — or knowledge — on generalization. Such procedural knowledge includes human decisions on which techniques are applied to actually generalize map features, the sequence in which these techniques are applied, and what tolerance values, or parameters, are used. Until researchers working in the spatial sciences have access to such procedural, and other forms of, knowledge, a fully automated approach to cartographic generalization seems unlikely. It appears to be timely for those in cartography and geographic information systems to turn their attention to issues of geographical and cartographical knowledge acquisition. This chapter will discuss general issues of knowledge acquisition. This is followed by a review of potential sources of both geometric and procedural knowledge involving existing rule bases, geocoding experiments, and user-interface design.

12.2 Knowledge acquisition

In knowledge acquisition for cartographic processes, including map design, type placement, projection selection, and generalization, a critical decision is in the selection of the **knowledge engineer**. The knowledge engineer is the chief architect, of course, of a given expert system, such as that for cartographic generalization. The specific task of building an expert system includes the process of information gathering, domain familiarization, analysis, and design efforts. More importantly, the accumulated knowledge must be translated into code, tested, and refined (McGraw and Harbison-Briggs, 1989). The knowledge engineer, the individual responsible for structuring the expert system, is also called the system designer, senior engineer, computer scientist, cognitive psychologist, systems engineer, or even programmer. Knowledge acquisition, also called knowledge extraction, and knowledge elicitation, then, is the process of transforming 'problem-solving expertise' from a knowledge source to a program (McGraw and Harbison-Briggs, 1989, p. 8). While the knowledge engineer designs the expert system and is responsible for acquiring knowledge, it is the **domain expert** who provides the actual knowledge. Thus, for the purposes of cartography, the domain engineer might be a USGS or Census Bureau cartographer.

Several researchers, including McGraw and Harbison-Briggs (1989), have suggested several methods for acquiring knowledge, including: (1) interviewing experts, (2) learning by being told, and (3) learning by observation. **Interviewing** involves the knowledge engineer meeting with and extracting knowledge from the domain expert. In **learning by being told**, the expert 'is responsible for expressing and refining the knowledge' while the knowledge engineer handles the design work (McGraw and Harbison-Briggs, 1989, p. 9). The third approach, **learning by observation**, is more complicated. Here, the expert is allowed to interact with sample problems or case studies, or even use previous case histories. In the field of cartography and GIS, important initial decisions will involve how, exactly, do we acquire cartographic knowledge, or, specifically, which of these techniques do we use?

12.3 Knowledge types

Several classifications of geographical/cartographical knowledge exist (Buttenfield and McMaster, 1991). Armstrong (1991) suggested three: geometrical knowledge, structural knowledge, and procedural knowledge. Geometrical knowledge involves feature descriptions on location and density; structural knowledge involves the intrinsic expertise of the scientist, such as soils scientist or demographer; procedural knowledge involves the necessary operations and sequencing. While Chang and McMaster (1993) agree with these, they also add semantic knowledge, which is often associated with the generating processes of a feature. Soils, for instance, vary considerably based on their geomorphic, generating, processes. Those in computer science, such as McGraw and Harbison-Briggs (1989), provide detailed explanations of knowledge types, albeit not from a geographical perspective.

12.3.1 Geometric knowledge from existing rule bases

Rules for the production of maps have been developed for decades. Birdseye, in a 1928 United States Geological Survey manual, discusses issues of generalization:

> . . . it is necessary to understand the uses and meaning of the terms "detail" and "generalization" and their application to Geological Survey maps. These terms describe relative conditions that are opposite or complementary. 'Detail' implies a refined treatment and suggests literal mapping; "generalization" signifies a broad treatment and involves an abridgement. The use of the terms may be further defined by an example. A map drawn on a scale of 1:24,000 should

> represent a region in detail as compared with a map of the same region drawn on a scale of 1:48,000, on which the representation must be confined to a broad generalization of the same features (Birdseye, 1928, p. 183).

Birdseye provides specific details for generalization, such as:

> . . . standard for the inking of the conventional country-house symbol on the field scales specified: 1:96,000, 150 feet square; 1:48,000, 75 feet; 1:31,680, 50 feet; 1:24,000, 25 feet. These dimensions, however, should be increased or decreased in congested places, if the legibility is correspondingly improved. Large structures which, when plotted to scale, exceed the size of the ordinary symbol should be shown with their individual plan outlines (Birdseye, 1928, p. 183).

or, even more specifically,

> Distances between houses are to be understood as from center to center, and length of street blocks as between building lines and not from center to center of streets.
>
> 1:48,000 scale field work for publication on 1:62,500.
>
>> Houses, where evenly or nearly evenly distributed, should be blocked where the distance between them is less than 100 feet — for example, in a 500-foot street block size houses evenly distributed should be blocked, but five houses or less should be shown separately.
>
> 1:24,000 or 1:31,680 scale field work for publication on either scale.
>
>> Houses should be blocked where the distance between them is less than 50 feet — for example, in a 500-foot street block 11 houses evenly distributed should be blocked, but 10 houses or less should be shown separately.
>
> 1:24,000 scale field work for publication on 1:62,500
>
>> Houses less than 100 feet apart should be blocked.
>
> 1:96,000 scale field work for publication on 1:125,000.
>
>> Houses less than 200 feet apart should be blocked (Birdseye, 1928, p. 231).

It is clear that thousands of such general rules may be culled from National Mapping Agency (NMA) reports and manuals. Other, more current, sets of compilation/production/generalization rules for the United States Defense Mapping Agency and United States Geological Survey have been reported in McMaster (1991) and Mark (1991). One urgent issue in developing a comprehensive research agenda for generalization should be a survey and documentation of existing knowledge in NMA manuals and publications.

12.3.2 Geometric knowledge from the encoding process

An experiment was conducted in order to extract geometric knowledge from digitized lines. This knowledge is based on, in part, an analysis of digitizer consistency since digitized linear information is theoretically sampled in the encoding process. Is there some aspect of geometry encoded in this process that might assist in the generalization of digital lines?

For much geocoded linear information, the spacing of points using a digitizing tablet is contingent upon the characteristics of the line and the speed with which the cursor is moved. Along a sinuous line, an operator would move the cursor slowly in order to capture the major characteristics of the curves. Alternatively, the operator would move the cursor rapidly along a smooth or straight coastline. Furthermore, Traylor (1979) has shown that digitizer operators change as they become more experienced. This might result in differences between the early digitized lines and those lines digitized near the end of data acquisition.

A technique was developed to acquire geometric knowledge for 31 digitized lines using two characteristics of the digitized line segments: density and direction. The specific directional characteristics of the lines were composed of many vectors at a very large scale. While the general trend of the line might very well be east–west, an enlargement of this line would reveal a great number of small angles not necessarily having an east–west trend. Is this information useful, in some way, in gathering knowledge of the feature?

Distance–direction matrix

A first step in acquiring the geometrical knowledge involved developing a series of distance–direction matrices. These matrices were based upon the nature of the digitizer sampling characteristics, including both the resolution of the digitizer (0.005 in.) and the Cartesian coordinate system. Given the x,y coordinate of the origin of a sampling vector, the motion of the cursor describes a vector in a limited number of directions and distances. After an inspection of several enlarged lines it became apparent that a high percentage of the vectors in the experimental lines were less than one-hundredth of an inch in length. This led to the development of a 28 point sampling matrix based on the 0.005 in. resolution of the digitizer (Figure 12.1). Twenty-four of the possible vectors were tabulated separately while the few longer vectors were tabulated in four additional matrix cells (labelled Q1–Q4 in Figure 12.1). A counter assigned to each cell was incremented every time the vector was directed to that cell. Figure 12.1 illustrates five such matrices for a sample line. For instance, the first matrix represents a vector movement of one resolution unit in the positive x-direction while the second matrix represents a vector movement of one digitizer resolution unit in both the positive x and negative y directions. The third matrix illustrates a longer vector that is recorded in the Q1 matrix cell.

A summary matrix was calculated for each of the 31 lines. The data for one such matrix are represented in Figure 12.2. In this instance, the vector was directed to the position (0.0, −0.005) 191 times, and to the position (−0.01, −0.01) only twice. The cell with 191 vector counts was assigned a rank of 1, while the cell with only two vector counts was assigned a rank of 28. These data were converted to ranks due to sampling limitations based on the discrete nature of the matrix. Since each digitizer grid intersection actually represents an area, not a discrete point, where coordinates are sampled, a digitizer records a coordinate to the nearest intersection.

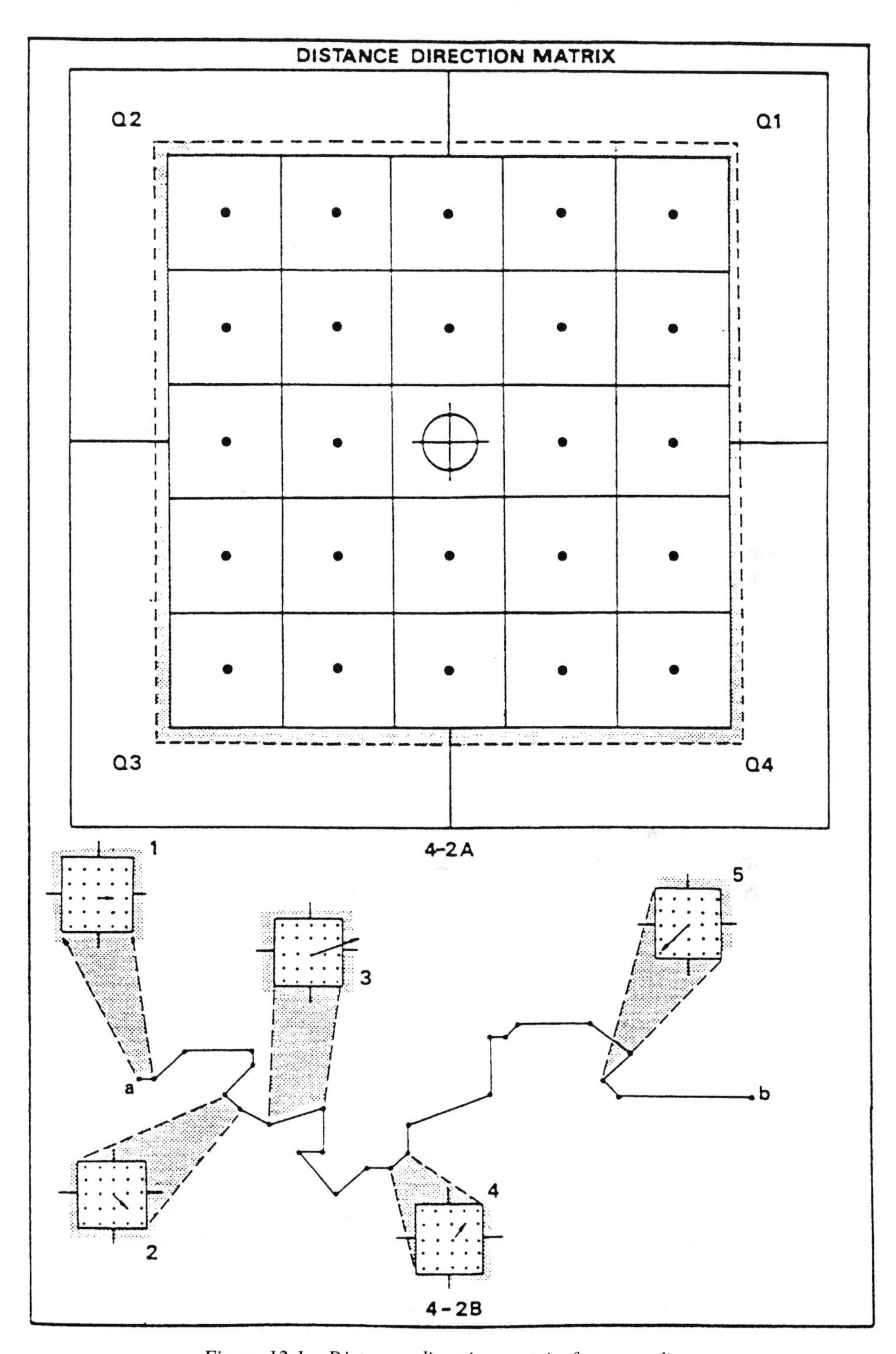

Figure 12.1 Distance–direction matrix for geocoding.

7						14
	6	13	33	11	3	
	20	61	86	36	14	
	91	191	0	98	50	
	24	77	162	44	13	
	9	15	59	9	2	
5						7

23.5	25	18.5	12	20	27	
	14	7	5	11	16.3	
	4	1		3	9	
	13	6	2	10	18.5	
	21.5	15	8	21.5	28	
26						23.5

Figure 12.2 Distance–direction matrix. The top matrix illustrates the raw counts of vectors directed to each cell. The bottom matrix represents the ranked version of these raw counts.

The result of these calculations produced 31 ranked matrices, one for each of the lines. Next, an average ranked matrix for all lines was calculated by summing the raw data for each cell for each of the lines and dividing the cell total by 31. This summed matrix was also ranked. The entire procedure resulted in 31 ranked matrices (one for each line) and a single summed and ranked matrix, created from an amalgamation of the individual matrices.

Geomorphic/geometric properties

The experiment involved determining if some type of structure inherent in the encoding process would provide useful knowledge for automated generalization, in particular simplification. In selecting the 31 lines an attempt was made to select a broad range of lines with a variety of geomorphic properties. To see potential similarities/dissimilarities in these cartographic data, a technique was developed for grouping the lines. Perhaps the most significant characteristic individuals use in differentiating lines is the specific shape. Furthermore, various shapes are distinguished by the total directionality, or number of different vector changes along the line. For instance, multi-directional lines, or lines with many orientations, are easily distinguished from lines with only one primary direction. The number of these directions is a function of the line's angularity. A straight coastline has a much different shape than a highly sinuous high-altitude stream or a gently meandering mature river. Individuals seem to judge the shape of the line on two criteria: the directionality of the line and the basic sinuosity of the line. Both of these are a function of the individual vector angularity.

Using individual vector angularity, as defined by the distance–direction matrices, an attempt was made to array, or group, the 31 lines into some logical format that exhibited some geomorphological, or geometrical, structure. A three-dimensional view of these distance–direction matrices provides insight into differences between two lines (Figures 12.3 and 12.4). Note, in particular, the difference between a line such as Honylake and a line such as Fairview. The surface representing the Fairview matrix clearly indicates an emphasis in a positive x-direction which, in fact, exactly matches the digitization for such a line: primarily in one direction. A line such as Honylake, conversely, is a closed polygon or a complete circle as digitized. Here, the matrix is almost symmetrical with a slight emphasis on the right (right represents a positive x-direction and is positioned at the bottom of the surface). The generated surface for a line such as Chaqcan, an intermediate line, is still balanced, but still somewhat shifted to the right. Could such knowledge be used for determining potential conflict with other features, or for establishing tolerance thresholds for simplification and smoothing? Acquisition of such geometrical knowledge from the digitizing process is just one potential source.

As a further experiment, a grouping of the matrices (lines) was performed using the common multi-variate technique of cluster analysis. This is a method of classifying objects into homogeneous groupings, based on their similarities. Clustering is accomplished by calculating a measure of similarity between every pair of objects, placing together the objects with the highest mutual similarity. The structure of the 31 lines was evaluated using the cluster analysis program NTSYS (Numeric Taxonomic System). Before applying this program, all data were standardized by converting the 28 cells of the raw data matrix to percentages.

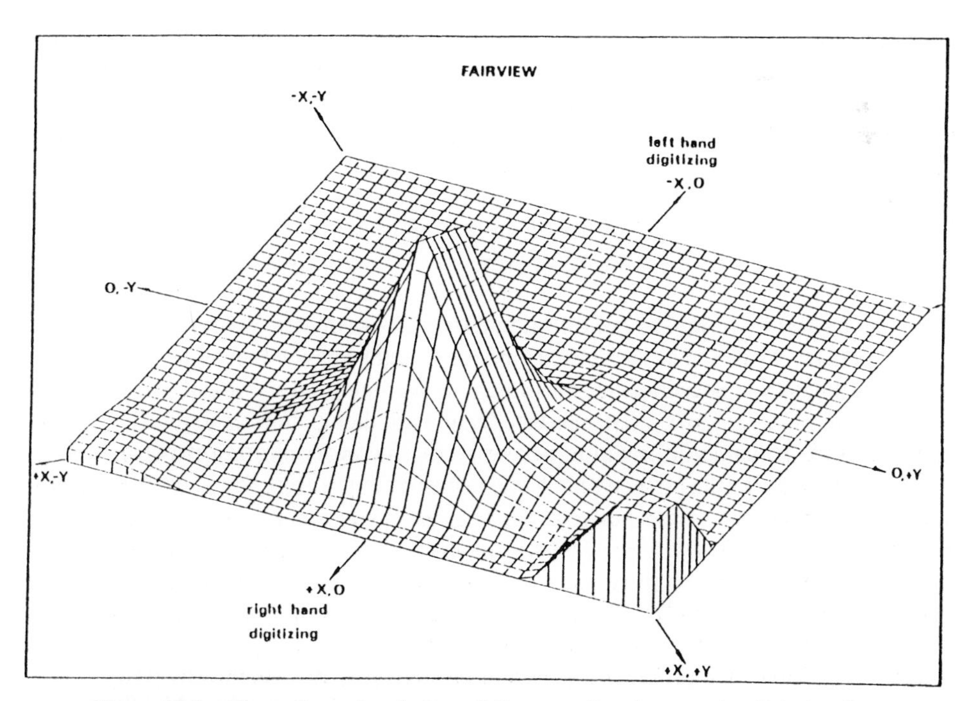

Figure 12.3 Three-dimensional view of distance–direction matrix: Fairview line.

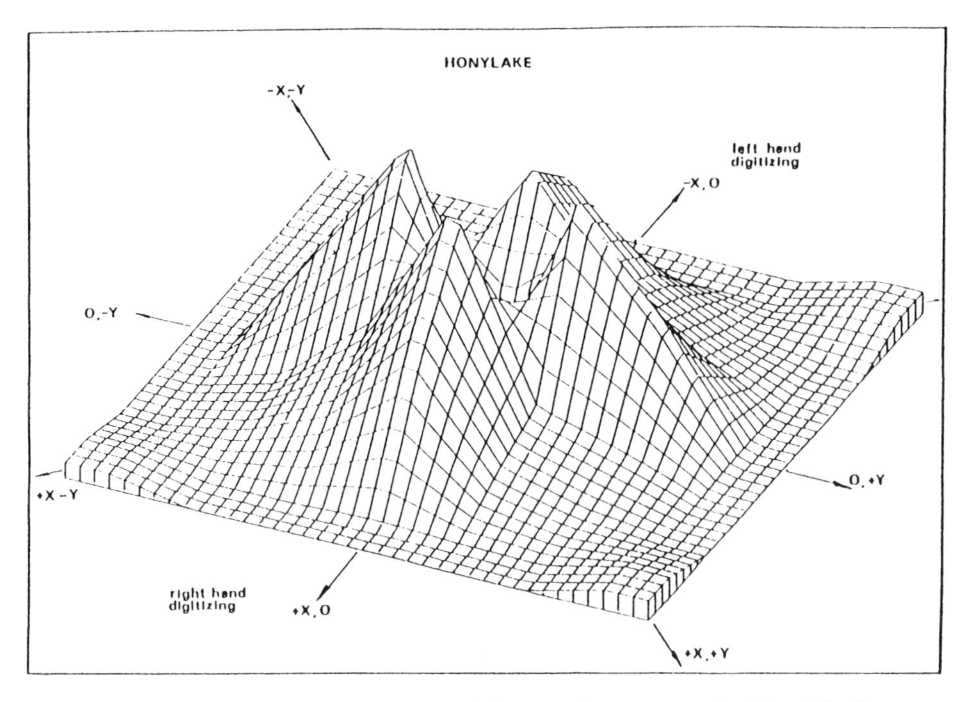

Figure 12.4 Three-dimensional view of distance–direction matrix: Honylake line.

Results of the cluster analysis

An analysis of the dendogram resulting from the cluster analysis suggested that the lines could be broken into three groups (Figure 12.5). The first cluster of lines, those lines that are more simple in nature, may be found on the left-hand side of Figure 12.5. These lines are relatively smooth and do not exhibit the sinuosity of cluster II. In fact, aside from some minor perturbations, line number 26 (Fairview) is essentially a straight line section of the southern coastline of Lake Erie. The second cluster is located in the center one-third of the diagram. These lines are quite sinuous, yet their primary trend is in one direction. The third cluster is positioned on the right-hand one-third of the figure and represents those lines that, for the most part, have a generally multi-directional trend. Five of the lines, numbers 17, 25, 27, 28, and 29, are almost closed polygons. All nine of these lines have at least three major directional trends. As a general classification method, the three major clusters may be labelled smooth straight lines, sinuous lines, and multi-directional lines. Again, within the constraints of the data and methodology, this technique provides for a generalized characterization of digital line data. Such methods should provide valuable information necessary for the generalization process. Similar work in evaluating the structure of cartographic lines has been completed by Buttenfield (1989).

12.4 User-interface design

Several researchers have proposed the design of user interfaces specifically designed for generalization and, in particular, for procedural knowledge acquisition (McMaster and

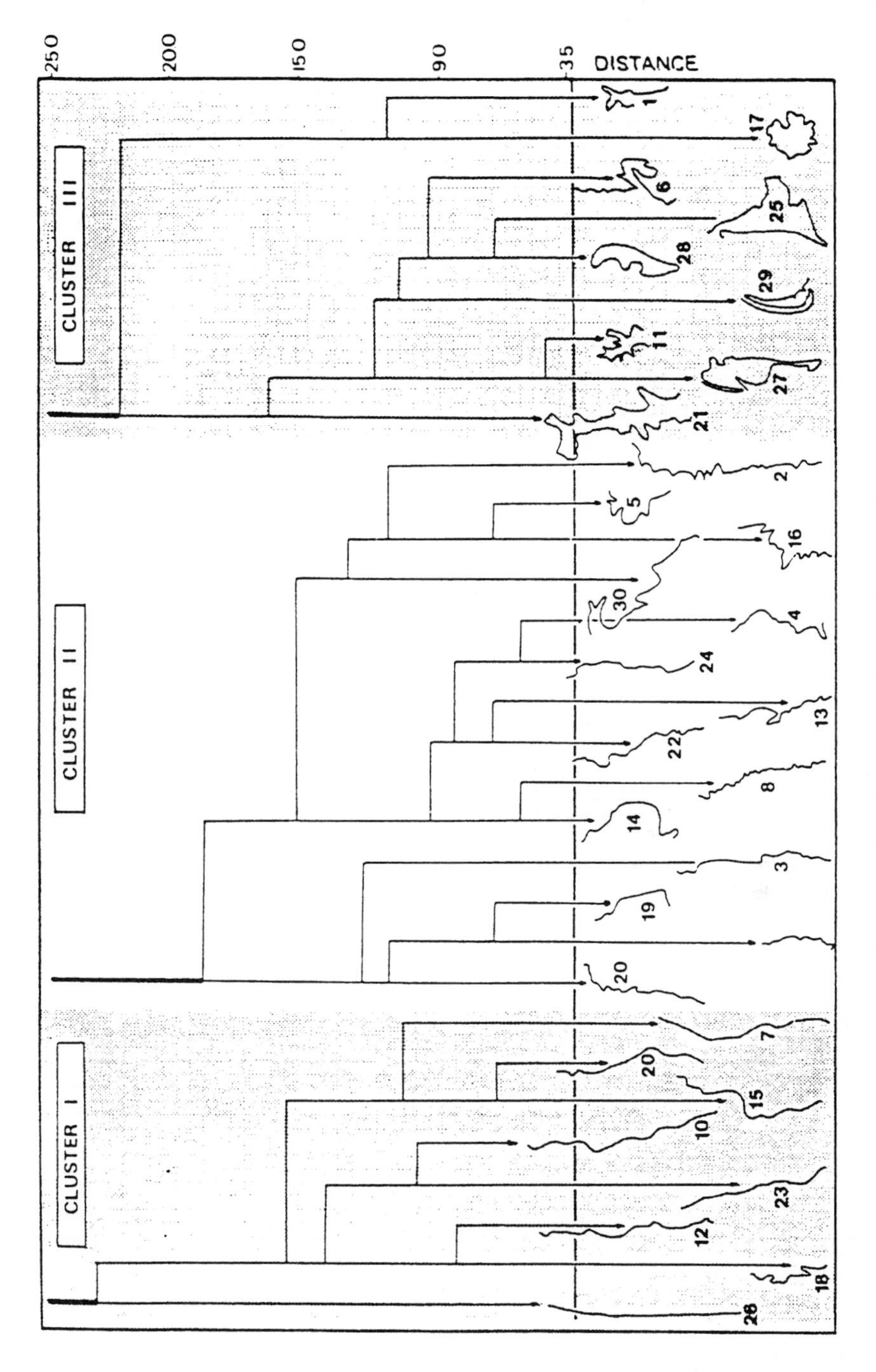

Figure 12.5 Results of cluster analysis of distance–direction matrix.

Mark, 1991; Lee, 1992; Chang and McMaster, 1993). Additionally, Weibel, in discussing the concept of amplified intelligence, identifies the need for a powerful user interface 'based on the design paradigm of direct manipulation and visual feedback' (Weibel, 1991, p. 178). McMaster and Mark (1991) provide specific objectives for a user interface, including:

1. Provide the user with a comprehensive set of generalization operators that will allow the user to manipulate the map image.
2. Provide a comprehensive set of tools that will assist in identifying map features.
3. Provide the user with assistance in selecting tolerances and parameters for individual operators and algorithms.
4. Provide the user with warnings when an illogical selection is made.
5. Provide a 'trace' or accounting of feature generalization.
6. Provide the user with hypermedia-based documentation and diagrams.
7. Provide the user, when possible, with a measure of success.
8. Provide the user with features on the map in need of generalization, or what may be called generalization 'hot spots'. Such regions where the density of features is high may be located by the conditions of generalization identified by McMaster and Shea (1992). Such conditions are necessarily evaluated through the application of measures that act as descriptors of the geometry of individual features and assess the spatial relationships among the combined features. Zhan and Mark (1993) have discussed a recent project for detecting conflict resolution in generalization.

In Figure 12.6, such an idealized interface is proposed. For the purposes of procedural knowledge acquisition, the concept of a user log is crucial (number 5 above). As the experienced cartographer or domain expert works with the map image, a detailed log of

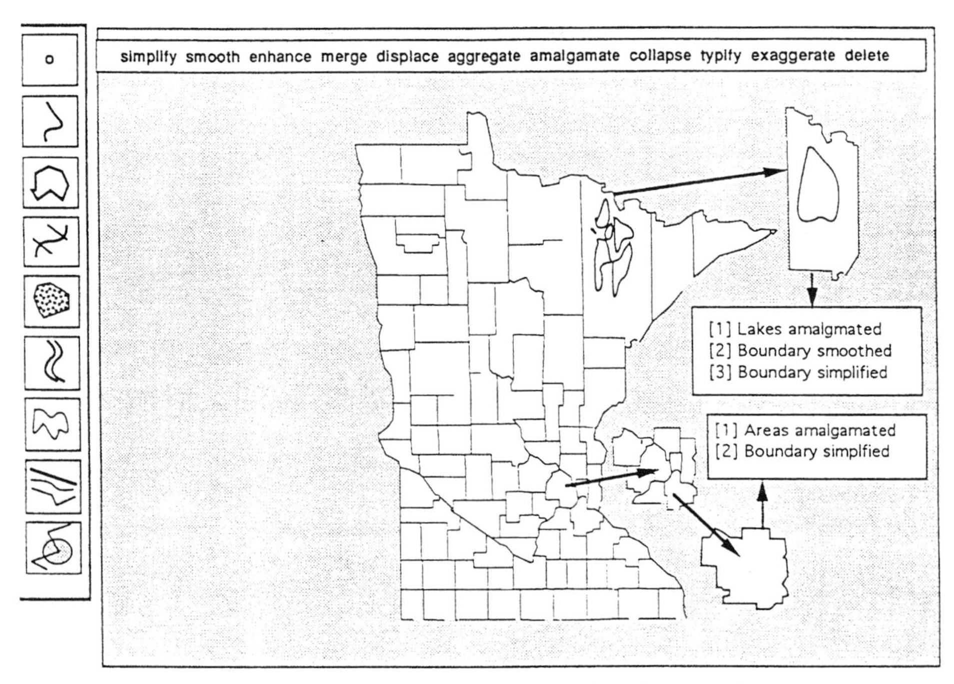

Figure 12.6 Prototype 'idealized' user interface for generalization.

activity is maintained. In the example depicted, lake boundaries are amalgamated, and the resultant regionalized boundary is smoothed and simplified. In the seven-county Minneapolis–St. Paul metropolitan region, the counties are amalgamated and boundaries simplified. It is expected that, through a careful examination of the logs from such sessions, specific, feature-based data on the generalization operators and parameters — fundamental procedural knowledge — may be obtained.

12.4.1 User interface for procedural knowledge acquisition

Chang and McMaster recently reported on such an interface, designed for procedural knowledge acquisition, at the University of Minnesota (Chang and McMaster, 1993). Using a Sun SPARCstation, a user interface (designed using SUNPhigs) has been developed. The basis for the user interface is a set of multiple windows that allows the user to experiment with different simplification and smoothing algorithms (McMaster, 1989). For each of the four windows created, a separate simplification algorithm is available. The user, through direct manipulation of a series of sliding bars for each window, may quickly change the scale, position of the feature, and tolerance value. It is also possible, for each of the simplification windows, to pull down a procedure window that allows: (1) overlay of the original feature, (2) smoothing of the feature using a series of smoothing algorithms, and (3) animation. In the animated sequence, the feature is simplified from the densest set of points to a caricaturized version in real time.

For instance, using window 1 a user could (1) simplify a feature using the Lang algorithm, (2) smooth the feature with a five-point moving average, and (3) overlay the original for comparative purposes. The system is now being improved with the addition of geometrical measures, such as change in angularity, and the option of stopping the animation at any time. Future development will incorporate additional generalization operators. Eventually, the user interface will be used with the generalization test data set, developed by the NCGIA, to gain procedural knowledge on generalization from trained professional cartographers or 'domain engineers'.

Figure 12.7 depicts the generalization interface using a North American database. This Map Generalization System (MGS) provides the user with four windows, each with one simplification routine. The four routines include: Lang algorithm (upper left), Douglas algorithm (upper right), Reumann algorithm (lower left), and VectGen (lower right). Descriptions of these algorithms may be found in McMaster (1987a) and are geometrically evaluated in McMaster (1987b). On this illustration, the same tolerance value (0.730 units) has been applied to each algorithm. Associated with each window — located to the right — are a set of slider bars for controlling the size of the map image and the level of simplification. The function of each slider bar includes:

1. manipulating the tolerance value,
2. enlarging/reducing the image (Zoom),
3. relocating the x-center,
4. relocating the y-center, and
5. a ratio value for enhancement (not yet functional).

For multi-parameter algorithms, it is necessary to include several slider bars to control the algorithm. For instance, the Lang algorithm requires two controls, one for the actual distance tolerance value (Tolere), and one for the number of points to look ahead (Pnt Ad).

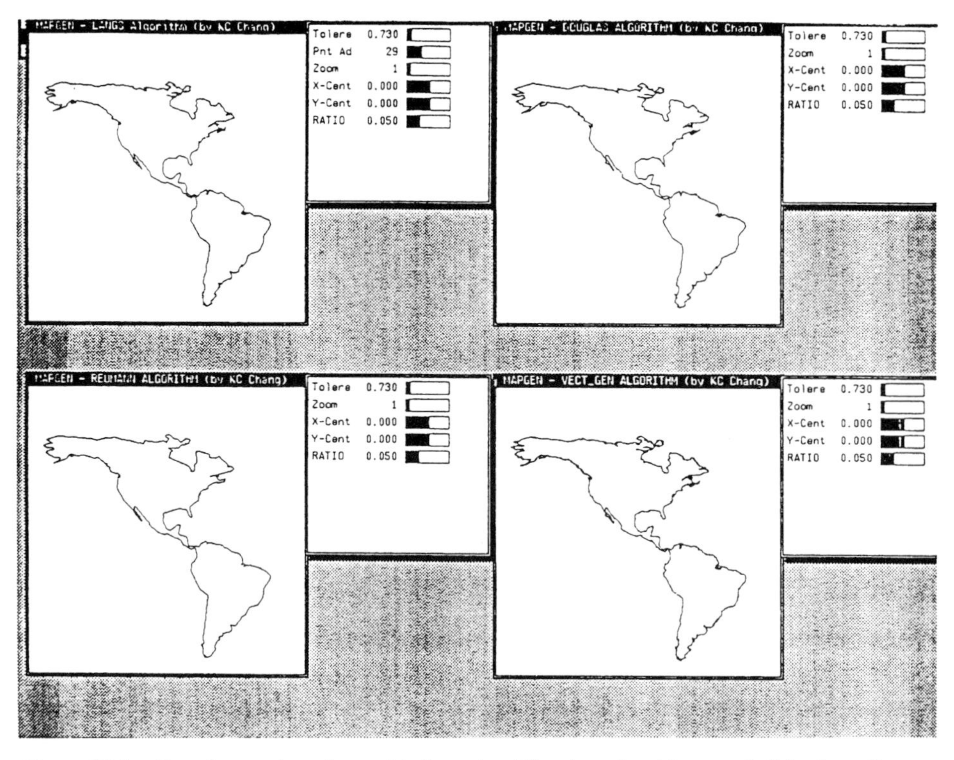

Figure 12.7 Sample user interface with four simplification algorithms and slider bars for controlling level of simplification.

Figure 12.8 illustrates the Zoom function, where the detailed coastline of Central America is illustrated. In Figure 12.9, where each of the four algorithms has been reduced to 0.883 units, the different results of the algorithm may be seen. Using the same tolerance value, the Douglas and Reumann algorithms behave much differently than the Lang and VectGen algorithms.

For each of the four windows, an additional menu bar may be pulled down as illustrated in Figure 12.10. This bar enables the following functions:

1. overlay of the original line,
2. animation,
3. smoothing using a B-spline,
4. smoothing using a five-point weighted moving average,
5. a simple enhancement routine,
6. changing of the base map,
7. QUIT Map Generalization System.

As noted in Figure 12.11, the user of the system may compare the simplified and smoothed versions for each of the four examples. This particular example depicts the application of a five-point moving average. As the line is simplified, through direct manipulation of the slider bar, the smoothed version is regenerated in real time. The original digitized feature can also be overlaid on the simplified and smoothed versions. Figure 12.12 depicts the selection of the second smoothing option, the B-spline, and

Figure 12.8 Enlargement of Central America using Zoom.

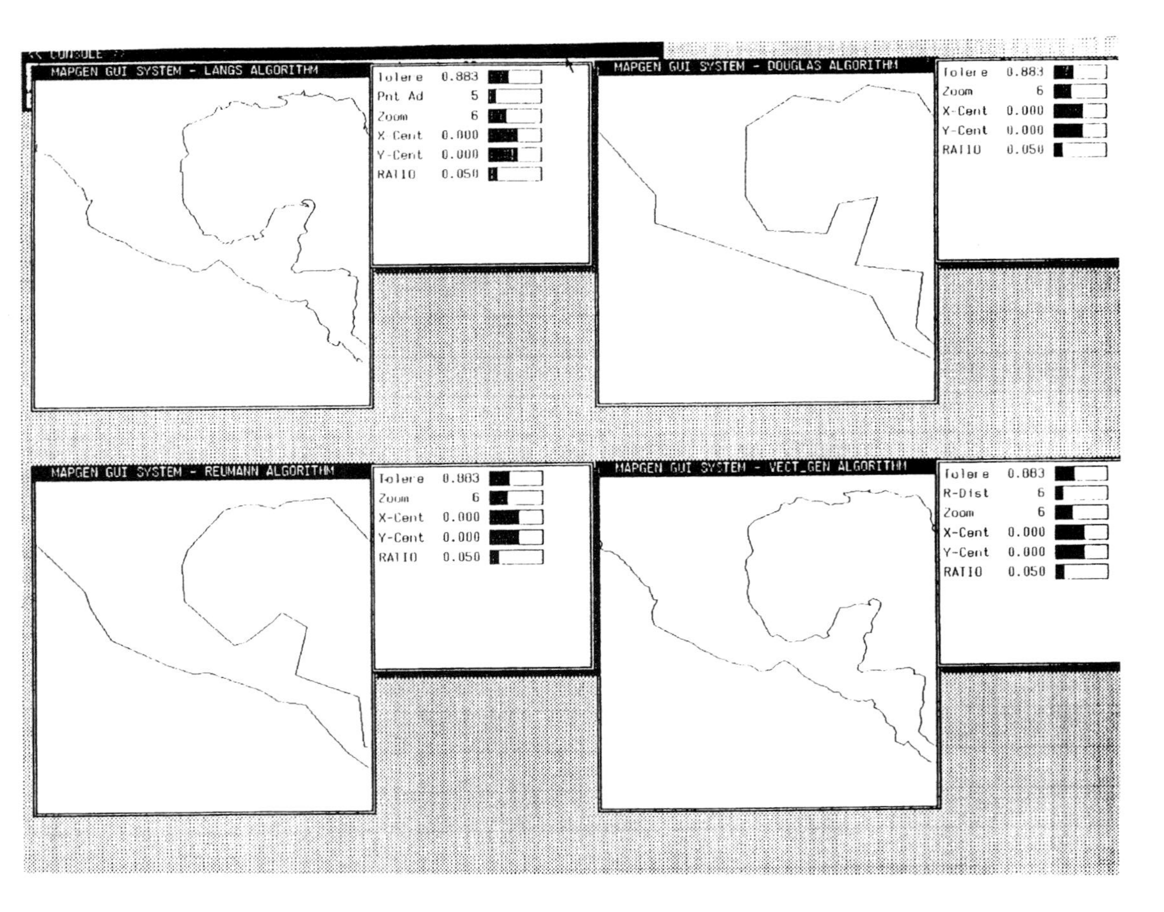

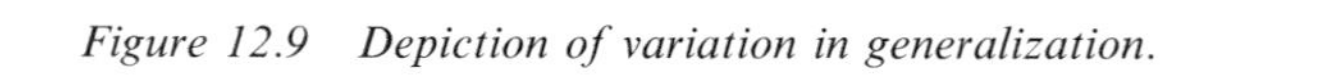
Figure 12.9 Depiction of variation in generalization.

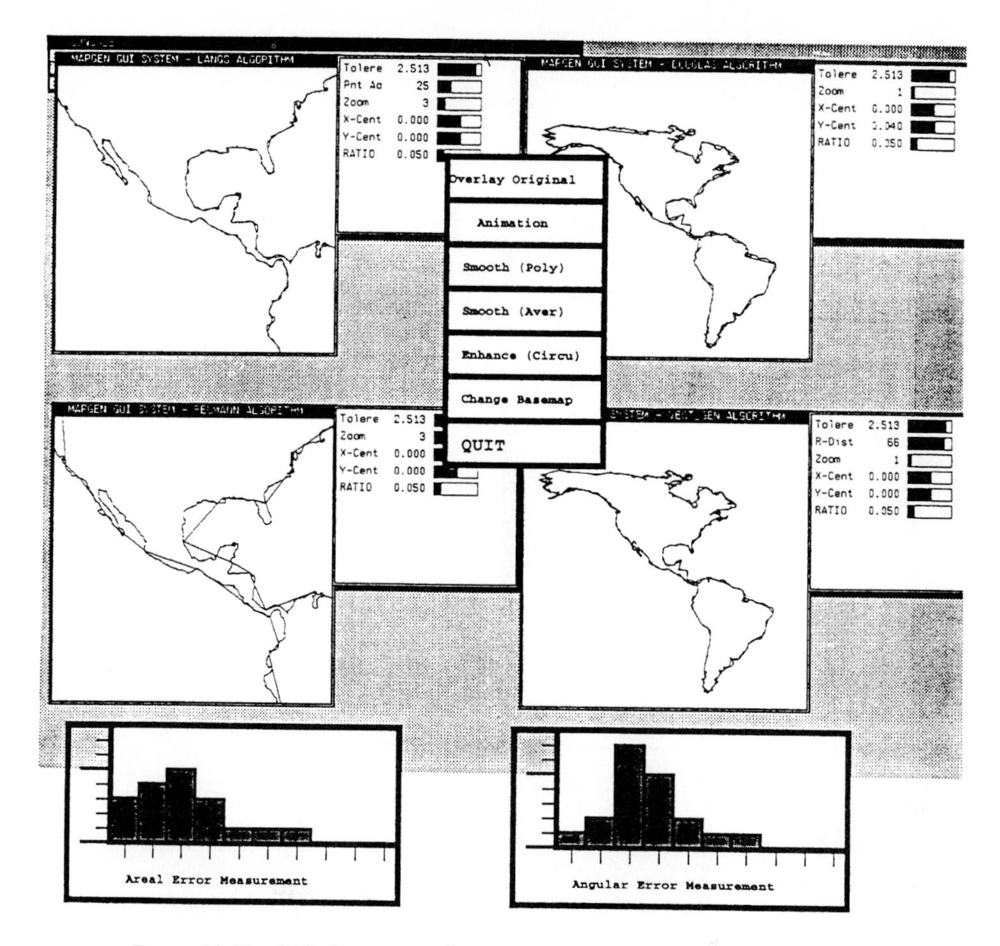

Figure 12.10 Pull-down menu for animation, smoothing, and enhancement.

compares this to the simplified version. As with the moving-average smoothing, direct manipulation of the simplification tolerance results in a real-time recalculation of the spline.

Future work with the interface will involve four tasks: (1) the addition of logging capability, (2) the addition of measurement capability, (3) adding new generalization operators, and (4) a redesign of the overall interface. Since the purpose of the Map Generalization System is to acquire procedural knowledge as the user manipulates each feature, the 'procedures', or operations, must be recorded. As explained earlier in this chapter, this involves logging the operations, algorithms, and parameter, or parameters, that are being used for each feature. Subsequent quantitative and qualitative analyses might involve evaluating these logging records to look for common approaches to generalization amongst various cartographers. The system as currently designed only has capability for two operations — simplification and smoothing. Other operators — enhancement, exaggeration, and amalgamation — must be carefully tested and added to the system. Additionally, real-time feedback to the user on basic geometric characteristics such as percentage coordinates eliminated and areal/vector displacement of features must be available. Lastly, an improved design for the format of the interface is being

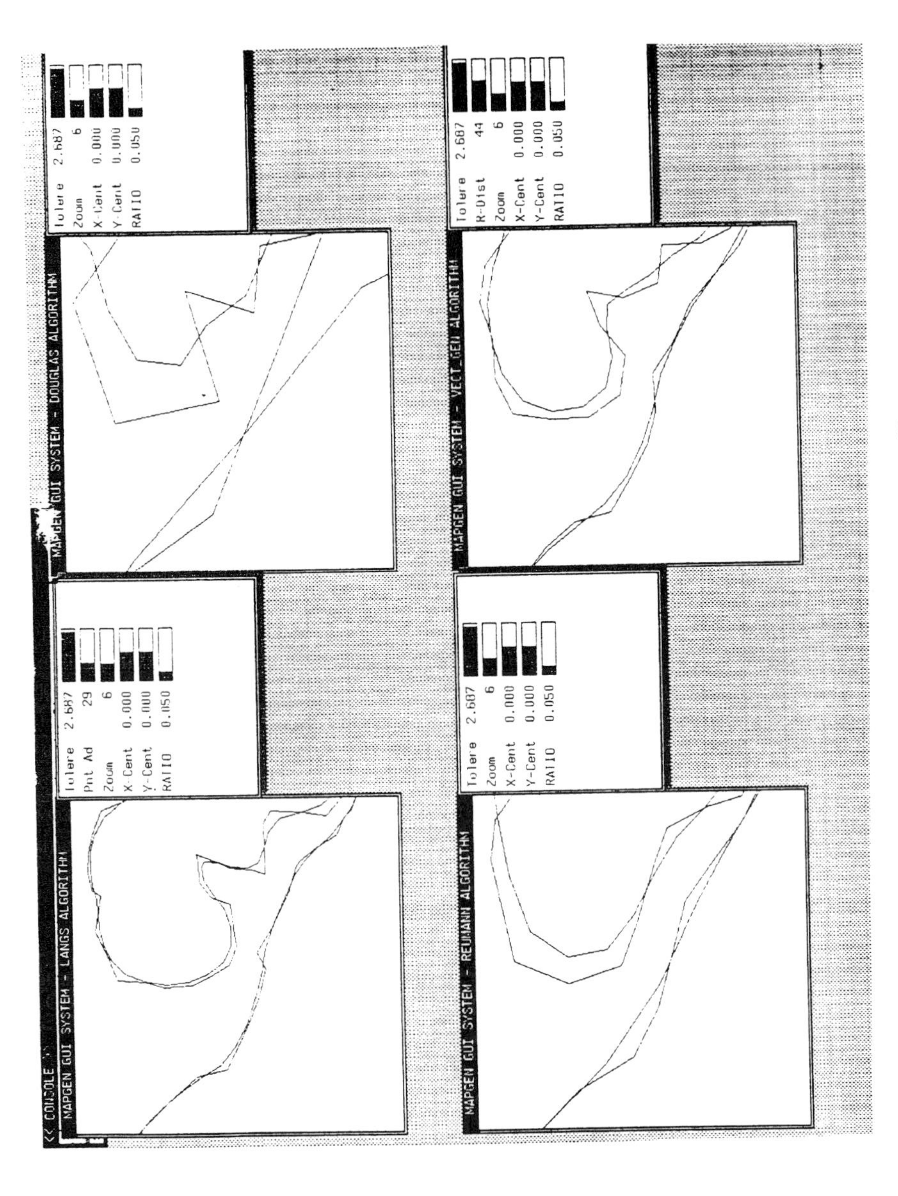

Figure 12.11 Moving-average smoothing.

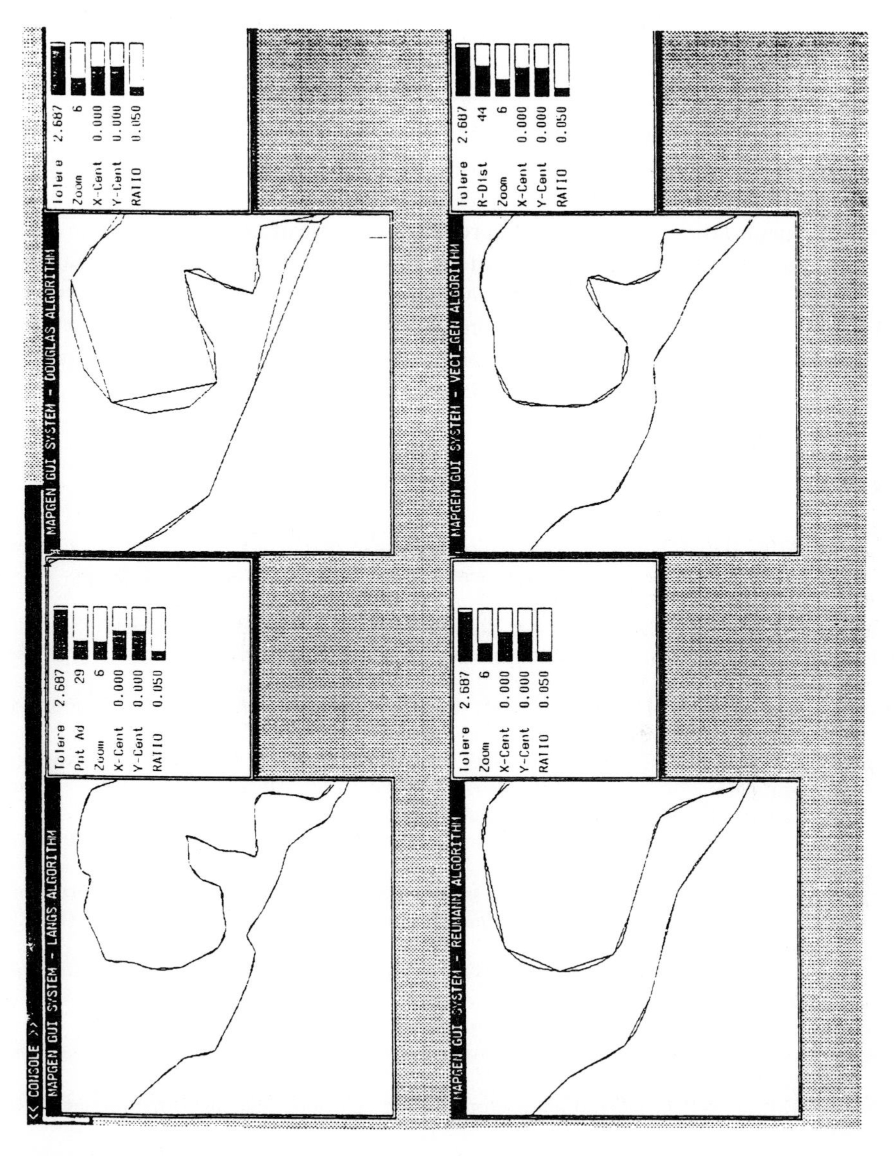

Figure 12.12 B-spline smoothing of four images.

considered. This will involve one large window with the map image, and several smaller windows with the target, smaller-scale, images. The simplification algorithms will be pulled down with a menu, as with the current structure for smoothing algorithms. As the user works with the larger-scale window, the same features will be modified in one smaller-scale window, which allows the user to 'recreate' a static image in the second smaller-scale window.

12.5 Summary

As GISs become more sophisticated, and the potential for quality cartographic output is finally realized, it will be necessary to pursue the application of expert systems. This will initially be in fairly small domains, such as typographic placement, symbolization selection, and generalization. The significant problem in the use of expert systems is in acquiring and formalizing the necessary cartographic knowledge. Capable domain and knowledge engineers must be identified, specific knowledge-acquisition techniques need to be developed, and well-thought-out simulations for complex cartographic processes should be designed. This chapter has presented several examples of both sources and techniques for knowledge extraction in the generalization process.

References

Armstrong, M.P., 1991, Knowledge classification and organization, in *Map Generalization: Making Rules for Knowledge Representation*, Buttenfield, B.P. and McMaster, R.B. (Eds), pp. 103–18, London: Longman.

Birdseye, 1928, USGS Production Manual, Reston, Virginia: United States Geological Survey.

Buttenfield, B.P., 1989, Scale-dependence and self-similarity in cartographic lines, *Cartographica*, **26**(1), 79–100.

Buttenfield, B.P. and McMaster, R.B., 1991, *Map Generalization: Making Rules for Knowledge Representation*, London: Longman.

Chang, H. and McMaster, R.B., 1993, Interface design and knowledge acquisition for cartographic generalization, in *Proceedings of the Eleventh International Symposium on Computer-Assisted Cartography, (Auto-Carto-11)*, Minneapolis, MN, pp. 187–96.

Lee, D., 1992, 'Cartographic generalization', unpublished Technical Report, Intergraph Corporation, Huntsville, AL, USA.

Mark, D.S, 1991, Object modelling and phenomenon-based generalization, in *Map Generalization: Making Rules for Knowledge Representation*, Buttenfield, B.P. and McMaster, R.B. (Eds), pp. 86–102, London: Longman.

McGraw, K.L. and Harbison-Briggs, K., 1989, *Knowledge Acquisition: Principles and Guidelines*, Englewood Cliffs, NJ: Prentice Hall.

McMaster, R.B., 1987a, Automated line generalization, *Cartographica*, **24**(2), 74–111.

McMaster, R.B., 1987b, The geometric properties of numerical simplification, *Geographical Analysis*, **19**(4), 330–46.

McMaster, R.B., 1989, The integration of simplification and smoothing algorithms in line generalization, *Cartographica*, **26**(1), 101–21.

McMaster, R.B., 1991, Conceptual frameworks for geographical knowledge, in *Map Generalization: Making Rules for Knowledge Representation*, Buttenfield, B.P. and McMaster, R.B. (Eds), pp. 21–39, London: Longman.

McMaster, R.B. and Mark, D.M., 1991, The design of a graphical user interface for knowledge acquisition in cartographic generalization, in *Proceedings GIS/LIS'91*, Atlanta, GA, pp. 311–20.

McMaster, R.B. and Shea, K. S., 1992, *Generalization in Digital Cartography*, Washington, DC: Association of American Geographers.

Traylor, C.T., 1979, 'The evaluation of a methodology to measure manual digitization error in cartographic data bases', unpublished Doctoral dissertation, University of Kansas.

Weibel, R., 1991, Amplified intelligence and rule-based systems, in *Map Generalization: Making Rules for Knowledge Representation*, Buttenfield, B. P. and McMaster, R.B. (Eds), pp. 172–86, London: Longman.

Zhan, F. and Mark, D.M., 1993, Conflict resolution in map generalization: a cognitive study, in *Proceedings of the Eleventh International Symposium on Computer-Assisted Cartography, (Auto-Carto-11)*, Minneapolis, MN, pp. 406–13.

SECTION V

Data Quality

13

The importance of quantifying the effects of generalization

Elsa Maria João

Department of Geography, London School of Economics, Houghton Street, London WC2A 2AE, UK

13.1 Introduction

One of the most important characteristics of Geographical Information Systems (GIS) is their ability to analyse spatial data. This can, however, be jeopardized by data transformations which detrimentally affect the data quality (Rhind and Clark, 1988). One such transformation that can alter the results of GIS map manipulations is the generalization of geographical data. Generalization can be described as a process by which the presence of geographical features within a map is reduced or modified in terms of their size, shape or numbers (Balodis, 1988). The end-product of a generalization process is therefore a derived dataset with less complex properties than those of the original dataset. For certain GIS applications, the maintenance of one large, detailed database could partly solve the problems caused by generalization. However, the ability to generate less-complex data from a detailed source dataset is still fundamental to GIS.

Within a GIS, generalization needs to be performed for three main reasons. The first, most obvious purpose is for *display*. The plotting of a map is one of the most common and useful outputs from a GIS map manipulation. Depending on the output scale and the detail of the source map, it might be necessary to generalize the data to improve depiction following the same principles used by manual cartographers. A map can often help communicate the results of a complex analysis in a clearer way than a set of tables and graphs. This can play an important role, e.g. in environmental impact studies where the environmental scientists carry out the analysis but often it is politicians that take the final decisions. Generalization carried out for display purposes is called *cartographic generalization.*

The second reason for the need to generalize within a GIS is strictly for *data reduction.* Driven either by financial or technological constraints, it may be necessary to generalize data in order to reduce the amount of data storage or processing time. The third main reason for generalizing in a GIS is for *analysis*. This can involve the use of generalization for homogenizing datasets which have different resolution or accuracy levels (see Weibel, Chapter 5). Most importantly, generalization as an analytical tool can be used to help understand at which scale spatial processes occur (Müller, 1991). Generalization carried out either for data reduction or for analysis is called *model generalization*. Some results of model generalization can also be displayed but, unlike cartographic generalization, the generalization was not done for the sake of graphic clarity.

Generalization, therefore, poses a dilemma to the GIS user. On the one hand it is necessary to generalize in order to improve the display quality of a map at a scale smaller than the one it was compiled from, or to allow analysis with different degrees of detail. However, on the other hand, generalization can potentially cause unintended transformations of the data that can alter the topology of geographical phenomena, and affect subsequent statistical or geometrical calculations. Normally, GIS users would want to minimize, control and quantify the effects of generalization on their results. Knowledge about the type and magnitude of generalization effects embedded within spatial datasets should therefore be deemed essential by any GIS user. Despite this, almost all the published literature describe only qualitatively the consequences of generalization. Other researchers (such as Beard, 1988) have measured generalization effects of specific manually generalized features, but solely in order to evaluate automatic generalization procedures. Blakemore (1983) and Goodchild (1980) investigated generalization effects directly but did not compare maps across different scales. Moreover, the qualitative descriptions have been mostly restricted to the domain of cartographic generalization. Thapa and Bossler (1992, p. 838), for example, point out that 'substantial' shifts can occur in actual ground terms due to cartographic generalization, but no figures are given.

This chapter discusses the importance of the generalization effects caused by manual and automated generalization, drawing on some of the findings of a quantitative study carried out by the author. João (1994) compared different source scale maps for two different study areas in order to measure the generalization effects embedded in the smaller-scale maps. GIS map manipulations were also carried out, using the same features taken from the different source scale maps, in order to determine the consequences of generalization on the results. The maps used had been manually generalized but, in addition, the largest scale map for each study area was generalized using the Douglas–Peucker algorithm and the results compared with the manually generalized maps. This chapter interprets the results of the study by João (1994) within the context of cartographic and model generalization. In order to illustrate the ideas presented in this chapter, an additional overlay operation was also carried out. This chapter concludes with a proposal by which, in future, automated generalization could increase its scope by encompassing the quantification and control of generalization effects.

13.2 Why we should still be concerned with the effects of cartographic generalization

Because the purpose of cartographic generalization is for display (i.e. it is a visual-oriented process), it can be deduced that some of its products are probably ill-fitted for certain analyses. 'One can not expect a database which was generalised using procedures for cartographic generalisation to be a reliable source of predictable quality for analysis, because manipulation by cartographic generalisation will introduce unpredictable errors through processes such as feature displacement' (Brassel and Weibel, 1988, p. 236). Data quality should determine which datasets are used for analysis. However, at present, there is widespread use of cartographically generalized products for analysis, independent of their quality. This is mainly because the vast majority of existing spatial databases have been digitized from existing map series (such as the ones produced by national mapping agencies) which were originally generalized for display reasons (Fisher, 1991). The quality of these databases remains largely unknown.

Most of the spatial databases derived from analogue map series do not have associated accuracy values. One reason for this is the difficulty of associating quality measures with digital data (see Smith and Rhind, 1993). Even when values are given they are often vague, relating, for example, to the dataset as a whole rather than to individual features or feature types, and they usually fail to refer specifically to the error associated with generalization. It has actually been found much easier to quote error values for recent map products that are minimally generalized. This has been the case with the Land-Line93, an Ordnance Survey product, derived from very large-scale maps (Smith and Rhind, 1993). Generalization error is very difficult to quantify (Thapa and Bossler, 1992) because the amount of error introduced depends on the type of feature, its proximity to other features and *when* the feature was inserted in the database. As new features (e.g. a new road) are updated, not only can they be drawn using different techniques, but they are often forced to adapt to the existing features in the database (the cartographer maintaining relative accuracy at the expense of absolute accuracy). As a consequence of this, the same type of feature can have different positional accuracy in different parts of a map according to the density of features and when the different parts of the map were updated.

The quantification of the effects of generalization can be especially problematic in that most GIS users do not produce their own data. Externally supplied data, obtained either in a digital or analogue format, will already have generalization effects embedded in them — the type and magnitude of these effects are often unspecified. An exception to this is the Australian Survey which indicates average displacements between different feature types on their 1:250 000 scale topographic database. For example, a typical displacement of up to 200 m can be observed in situations in which one road and one railway are almost coincident and the road must be moved to ensure clarity (Australian Survey, 1992). However, it is not stated which features or portions of features suffered displacements of this magnitude, if any at all.

Given the present situation, GIS users will often ignore the lack of accuracy of their digital data due to generalization. Because of this, generalization has been pointed out as one of the main causes as to why the magnitude of errors in current GIS databases can be larger than the errors within their analogue counterparts. According to Openshaw (1989) the magnitude of errors in spatial databases can surpass those introduced by traditional cartographic manipulations of paper maps due to the way GIS perform operations on cartographic data 'which traditionally would not have been done, or else performed only under special circumstances, because of the problems of scale, complexity, and feature generalisation that might be involved' (Openshaw, 1989, p. 263).

Most GIS users are therefore put in the position in which the only available digital products are the ones derived from traditional analogue map series, often of an unpredictable quality. This lack of error information about cartographically generalized maps would be of less importance if they were not used for analysis. Unfortunately, in practice, these products are widely used to support spatial queries for GIS applications. In order to evaluate the consequences of cartographically generalized maps on GIS operations, João (1994) carried out an extensive quantitative study. This study found that cartographic generalization could strongly affect basic measurements such as length. For example, some features' length changed as much as 23 per cent between the scales 1:50 000 and 1:500 000. More importantly, features were especially affected in terms of positional accuracy. Displacements of well-defined points (such as intersections between roads and railways) as large as 994 m were found between the scales 1:50 000 and 1:625 000. The results also showed that the magnitude of the generalization effects could be exacerbated in a typical GIS analysis. This is because for most GIS map manipulation operations any generalization effect can be compounded when two or more features are considered simultaneously.

In addition to the fact that cartographic generalization can strongly affect the quality of digital data, the unpredictable nature of this process makes it a more serious and difficult problem to tackle. The study by João (1994) found that the results of cartographic generalization were very variable according to feature classes but were even more unpredictable across individual features. It was often difficult to predict beforehand which of the feature classes would be most affected by generalization and for which of the different scales. For example, the feature class that presented the highest lateral shift at one scale was not necessarily the same as that having the highest value at another scale. This made it impossible to generate hard and fast cartographic generalization rules for individual feature classes. The variability and associated unpredictability of cartographic generalization is encapsulated in the notion that often manual generalization is governed by *best depiction* rather than by a *particular rule* (Geoff Johnson, Ordnance Survey, 1993, personal communication). The fact that cartographic generalization does not always produce predictable results, due to the extreme variability of the distribution of geographical phenomena, is a major difference between cartographic and model generalization.

The unpredictability of the effects of cartographic generalization reflects itself more seriously in the results of GIS map manipulations. In order to illustrate further the effects of generalization on a typical GIS map manipulation, a straightforward overlay operation was carried out to find out the length of a certain road which lay, respectively, within 100, 200 and 300 m of a river. Figure 13.1 shows the river (the Great Stour, that passes through the city of Canterbury in the south of England) and the road (the A28) used in the overlay operation.

The overlay was repeated using the same river and same road taken out of three different source scales: 1:50 000, 1:250 000 and 1:625 000. The data were obtained in a digital format from the Ordnance Survey of Great Britain. The 1:50 000 data were part of a trial digital product, but data from the other two scales are current digital cartographic data products sold by the Ordnance Survey. Table 13.1 shows the results of the overlay operation for the three distances when the features for the three different scales were used. It can be seen, for example, that the length of the road that lies within 100 m of the river Great Stour at the scale 1:250 000 is only 35 per cent of its length at the scale 1:50 000, while at the scale 1:625 000 it is only 52 per cent of its original length.

The results presented in Table 13.1 illustrate the issues mentioned above in connection with the effects of cartographic generalization. The fact that in the case of the 100 m buffer at the scale 1:250 000 only 35 per cent of the original road length remains, reinforces how large the effect of cartographic generalization can be. The reason for such a large effect is partly due to the geographic position of the two features — as the road and the river run parallel to each other (see Figure 13.1), even small displacements can cause a large impact on the results. The extent of cartographic generalization (such as the simplification, displacement and elimination of selected detail) is determined by the cartographer's aim to achieve a satisfactory representation of the landscape and this, in turn, is affected by the original landscape itself.

Moreover, the generalization effects do not necessarily increase progressively with the decrease of scale. In the case of the 100 m buffer, the overlay generated the smallest road

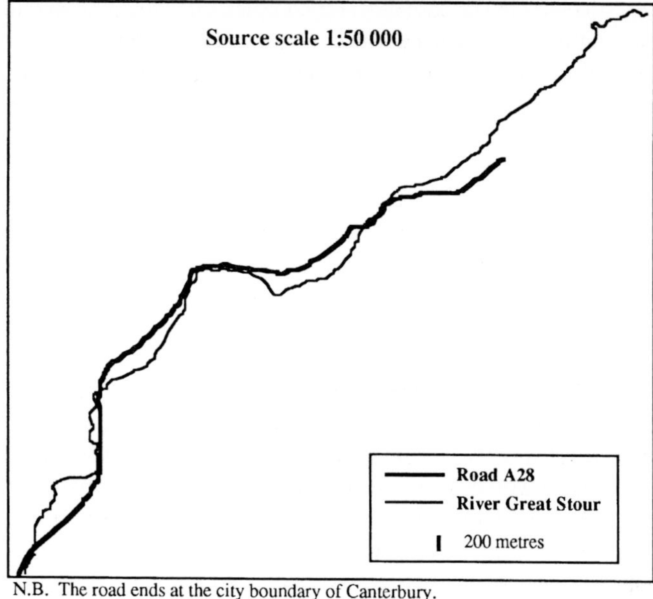

Figure 13.1 The river and the road used in the overlay operation.

Table 13.1 Results of the overlay using manually generalized features

Scale	Length of the road A28 within certain distance of the river Great Stour (m)		
	Within 100 m of the river	Within 200 m of the river	Within 300 m of the river
1:50 000	4548	7287	9658
1:250 000	1584 (35%)	5433 (75%)	8946 (93%)
1:625 000	2385 (52%)	3135 (43%)	4973 (52%)

length at the middle scale rather than at the smallest scale. Because of the unpredictability of the generalization effects, the way generalization influences GIS map manipulations is also not straightforward. Deriving data from a larger scale does not necessarily mean a reduction of generalization effects (cf. where model generalization can actually be used to increase accuracy — see Müller, 1991). With cartographic generalization some generalization rules used by cartographers might actually cause larger generalization effects at larger scales. Such is sometimes the case when a house is positioned between a road and a river. If, at a particular scale, a rule ensures that the house should be included, then this house might force the river and the road to move apart to allow room for the clear depiction of the house. However, if at a smaller scale the house no longer needs to be shown, then the river and the road might no longer need to be displaced.

The findings of this overlay operation corroborate the results of João (1994). They show both that errors associated with cartographic generalization can be large and unpredictable — especially in the case of more complex GIS map manipulations that use a large variety of features. As a consequence, it is particularly important to measure the impact that these errors are causing on GIS map manipulations, as long as cartographically generalized data continues to be used for analysis purposes. In the future — as digital data specifically tailored for analysis become increasingly accessible — the use of cartographically generalized products for analysis purposes will almost certainly diminish. Only when digital data generalized by model generalization (i.e. for analysis) are commonly available, might there be a less pressing need to attach accuracy values to cartographically generalized maps. The control of the data quality of *model* generalization will instead become increasingly important.

13.3 Model generalization and data quality

The importance of data quality can be encapsulated in the notion that information of unreliable quality can be worse than no information at all. For many GIS applications, users will only be able to evaluate the fitness for purpose of their data if they have access to specific quantitative measures of accuracy (Chrisman, 1991). Data can be considered as the most valuable asset of a GIS (Rhind, 1991) and so this concept of fitness for purpose within a GIS relates to the *multiple* applications which the data have the potential to be used for. In the situations in which data might be needed for purposes that had not been foreseen when the data were created, it is essential to have information on data quality to determine the suitability of the data for those new applications. The importance of enabling users to judge the fitness of data for their own use has been recognized by the US Spatial Data Transfer Standard (SDTS) which requires a data quality report coupled with every data transfer (Fegeas *et al.*, 1992).

Model generalization is usually done for analytical reasons and therefore, by definition, needs to keep close control of the impact of generalization on data quality. According to Brassel and Weibel (1988), model generalization aims at minimizing error, such as minimum average displacement, and so has to be done under parametric control. As more tools for model generalization start being developed (see Grünreich, Chapter 4), there will be a pressing need for a systematic analysis of the effects of model-oriented generalization.

In order to contrast the effects caused by model generalization with those caused by cartographic generalization, João (1994) generalized features with the commonly used Douglas–Peucker algorithm (Douglas and Peucker, 1973) and compared them with the same features generalized by manual cartographic methods. The Douglas–Peucker algorithm is an example of an algorithm which is specific to neither analysis nor display — it depends why and how it is being used. Because the algorithm was used with very small tolerances, no potential topological errors occurred (such as lines crossing back on themselves — see Müller, 1990; Visvalingam and Whyatt, 1990). This meant that, for the tolerances used, it conformed with model-oriented quality objectives (see Weibel, Chapter 5) and so can be considered a model-generalization tool.

João (1994) found that generalization effects were typically greater in manually generalized topographic maps than in those produced by the Douglas–Peucker algorithm. Automatic generalization retained length and angularity very well, and most importantly, displaced features much less. In other words, it caused much less distortion than manual generalization. This is because the Douglas–Peucker algorithm only filters the high-frequency components, causing a reduction in local detail of the lines without the more global displacement that the manually generalized lines often suffered. This is related to the way most line simplification algorithms work — they do not displace a line along its entire length, as even if the line loses the majority of its points, its first and last point will remain constant. These findings support the results of Beard (1987) who found that, for a thematic map depicting land and water areas, both positional and attribute errors were reduced by automatic generalization.

Model and cartographic generalization share the use of some common generalization tools, such as selection, simplification and smoothing. However, the tools exclusive to the use of cartographic generalization (such as enhancement, feature displacement and shape change) are the ones that are particularly responsible for the displacement and distortion of mapped features. João (1994) also found that because cartographically generalized features usually suffered more displacement than model generalized features (e.g. in terms of areal displacement as defined by McMaster, 1987), the results of GIS map manipulations were less affected by model generalization.

Table 13.2 illustrates how model generalization can affect the results of a typical GIS map manipulation to a lesser degree than cartographic generalization (compare with the results presented in Table 13.1). It shows the results of the same overlay operation as described in section 13.2, but this time using automatically generalized features. The road and the river at the scale 1:50 000 were generalized using the Douglas–Peucker algorithm. The level of generalization was determined by the number of crucial points (i.e. spurious points which did not add any extra detail were discounted) used to represent the road and the river at the scales 1:250 000 and 1:625 000.

It can be seen from Table 13.2 that although, in general, there is a constant increase in the generalization effects, these effects are much smaller than those for the manually generalized lines (cf. Table 13.1). However, there still remains some degree of unpredictability. In the case of the 300 m buffer, the result of the overlay at the smallest scale generated a slightly larger road length than at the largest scale. As in the case of the lines generalized manually, this was due to a combined effect of the reduction of length and the change of the relative position of the road and the river. The reduction by the algorithm of the number of points used to represent the lines caused a decrease of the road length but at the same time caused the sideways shift of sections of the road or the river (vector displacement as described by McMaster, 1987). It was this sideways shift which caused sections of the two lines to lie closer together, and therefore a lengthening of the road within 300 m of the river at scale 1:625 000.

Table 13.2 Results of the overlay using automatically generalized features

	Length of the road A28 within certain distance of the river Great Stour (m)		
Scale	Within 100 m of the river	Within 200 m of the river	Within 300 m of the river
1:50 000	4548	7287	9658
Equivalent to 1:250 000 (automatically generalized)	4488 (99%)	7281 (100%)	9626 (100%)
Equivalent to 1:625 000 (automatically generalized)	4461 (98%)	7122 (98%)	9822 (101%)

Despite the fact that it has been found that automated generalization can cause less generalization effects than manual generalization (Beard, 1988; João, 1994), the data quality control of algorithms such as the Douglas–Peucker is still very rudimentary. At present, most generalization algorithms lack mechanisms for the control of the quality of their output. Ideally, a GIS user would require help in the choice of the algorithm's tolerances so as to avoid, for example, generating lines which were excessively spiky or lines crossing back on themselves. So far, to a large extent, the control of the consequences of generalization on data quality has been invariably dissociated from the research into its automation. This is despite the fact that the automation of generalization would benefit from simultaneously scrutinizing the quality of its generalized products.

13.4 Increasing the scope of automated generalization by controlling its effects on data quality

Within a GIS, not only are the consequences of uncontrolled generalization more serious, but at the same time the quantification and control of generalization can be made easier by automation. Despite this, most of the effort in automating generalization has been dissociated from the control of generalization effects. As a consequence, most of the existing automated generalization tools cannot control the distortion they cause on the data, except for the elementary setting of the generalization tolerance (João, 1991). In particular, as model generalization advances, it is important that a system includes the minimization and quantification of unwanted generalization effects as an integral part of its model generalization tools.

To overcome this problem, a three-stage control process of generalization effects based on quality measures is proposed here. The first stage in this process would start by evaluating the need for generalizing. The control of generalization effects could then start even before the generalization process begins. GIS functions could give advice to the user about whether generalization is required or advisable — for example, by identifying conflicts (e.g. caused by overlapping or imperceptible features) using the approach suggested by Beard and Mackaness (1991). It is also important to define the limits of what is possible and sensible to attempt, and to educate the user about these limits. For example, there may be situations where it is inadvisable to combine a dataset at a particular scale with data from a very different scale, without generalizing the larger-scale data.

If it was confirmed that generalization was needed, then the second stage could control how, and how strongly, features were generalized, based on quality parameters. This could be carried out using the system suggested by João *et al.* (1993) in which the selection of the best available generalization procedure (in terms of algorithms and respective tolerances) was based on purpose, scale, data types and, most importantly, *quality requirements*. This improved automated generalization system would select generalization processes on the basis of the minimization of generalization effects as specified by the user. For example, the extent of generalization effects could be minimized by avoiding over-simplification of a line by the user specifying the maximum allowed length change or displacement.

The final stage of generalization control would take place after the generalization process was completed. This would involve a quantification and storage of the transformations caused by generalization. The quantification of generalization effects can be done at two different levels of detail. The simplest would entail storing only the quality constraints used in the generalization procedure (e.g. the maximum threshold value of allowed deviation between models — see Weibel, Chapter 5). A more detailed quantification of generalization effects would entail actually measuring what had happened. These *quantified* generalization effects could then be used to tag the data (so as to avoid misuse) or could be taken into account in GIS map manipulations based on error propagation models (Goodchild and Gopal, 1989; Openshaw *et al.*, 1991).

13.5 Conclusions

It is often argued that a dataset generalized for display purposes should not be used for analysis. However, this ignores the fact that, at present, they often are. If cartographically generalized products are the only type of digital data available, then GIS users have no other alternative than to use them. At the same time, if these cartographically generalized products lack quality values associated with them, it will be very difficult (or even impossible) for users to evaluate the acceptability of these datasets for analysis purposes.

This chapter has discussed how cartographic generalization effects are usually more serious than the effects caused by model generalization. The implications of this are two-fold. First, it emphasizes the importance at the moment of controlling and quantifying the effects of cartographic generalization on GIS map manipulations, while a large proportion of datasets used for analysis continue to be derived from cartographically generalized analogue maps. Second, it points out the pressing need for the development of more model generalization tools for generating datasets specifically suited to GIS analysis. As datasets derived from model generalization become increasingly available, it will, in turn, become a priority to evaluate and control their generalization effects.

Generalization is important to GIS, not only because of the drive towards more and better-automated tools but also due to the way generalization affects data quality. The development of automated tools and control of generalization effects have so far been dissociated to a large degree. In future, however, the control of generalization effects should be an extra parameter to be taken into account during the automation process. Research into the control of the quality of the products produced by generalization does not need therefore to be separated from the research into the automation process.

Independently of the development of a more intelligent system that would control generalization effects according to the preferences of users (João *et al.*, 1993), there is still scope for improving the information given to users about the type and magnitude of generalization contained in spatial datasets. If quality requirements are used to guide generalization, it should be possible to store these quality requirements as indicators of the quality of the generalized product. The information given by the Australian Survey (1992) in terms of the expected displacements for different feature types according to the number of features under conflict, sets the example that other mapping agencies should follow and enhance.

In addition, all digital datasets should be accompanied by a statement on data quality statistics that would indicate to the user the accuracy of the different features in different parts of the map. SDTS suggests three methods of quality reporting: textual narration, defined-quality attributes, and quality overlays (Fegeas *et al.*, 1992). For example, in order to help the user visualize generalization error in the database, 'grey boxes' could be drawn around heavily cartographically generalized areas to warn about excessive generalization effects and possibly motivate the user to obtain larger-scale maps (or to investigate the possibility of using data derived from model generalization).

In future, it might be the case that mapping agencies will commonly supply different datasets at the same source scale (say 1:250 000), but one which resulted from cartographic generalization and another that resulted from model generalization. These two distinct datasets would have different accuracy levels, bearing in mind the different types of uses that the data would have. However, the concept of fitness for purpose can only be evaluated and judged by the users themselves and not by the producer of the data (Chrisman, 1986). For most applications, this requires access to specific quantitative measures of accuracy rather than general qualitative statements. It is therefore fundamental that when automating generalization there is an associated quantification and control of the transformations suffered by the data.

Acknowledgements

Comments and suggestions from anonymous reviewers are gratefully acknowledged.

References

Australian Survey, 1992, *TOPO-250k Data User Guide*, Belconnen: Australian Surveying and Land Information Group.

Balodis, M., 1988, Generalisation, in Anson, R. (Ed.) *Basic Cartography for Students and Technicians*, 2, pp. 71–84, London: Elsevier, on behalf of the International Cartographic Association.

Beard, K., 1987, How to survive on a single detailed database, in *Proceedings of AUTO-CARTO 8*, Baltimore, MD, 30 March–2 April, pp. 211–20.

Beard, K., 1988, 'Multiple representations from a detailed database: a scheme for automated generalization', unpublished PhD thesis, The University of Wisconsin-Madison.

Beard, K. and Mackaness, W., 1991, Generalization operations and supporting structures, in *Proccedings of AUTO-CARTO 10*, Baltimore, MD, Vol. 6, pp. 29–45.

Blakemore, M., 1983, Generalization and error in spatial data bases, in *Proceedings AUTO-CARTO 6*, Falls Church, VA, 1983, pp. 313–22.

Brassel, K. and Weibel, R., 1988, A review and framework of automated map generalization, *International Journal of Geographical Information Systems*, **2**(3), pp. 229–44.

Chrisman, N., 1986, Obtaining information on quality of digital data, in *Proceedings of AUTO-CARTO London*, London, Vol. 1, pp. 350–8.

Chrisman, N., 1991, The error component in spatial data, in Maguire, D., Goodchild, M. and Rhind, D. (Eds) *Geographical Information Systems: Principles and Applications*, pp. 165–74. Harlow: Longman.

Douglas, D. and Peucker, T., 1973, Algorithms for the reduction of the number of points required to represent a digitised line or its caricature, *Canadian Cartographer*, **10**, 112–22.

Fegeas, R., Cascio, J. and Lazar, R., 1992, An overview of FIPS 173, the Spatial Data Transfer Standard, *Cartography and Geographic Information Systems*, **19**(5), 278–93.

Fisher, P., 1991, Spatial data sources and data problems, in Maguire, D., Goodchild, M. and Rhind, D. (Eds) *Geographical Information Systems: Principles and Applications*, pp. 175–89, Harlow: Longman.

Goodchild, M., 1980, The effects of generalisation in geographical data encoding, in Freeman, H. and Pieroni, G. (Eds) *Map Data Processing*, pp. 191–205, New York: Academic Press.

Goodchild, M. and Gopal, S. (Eds), 1989, *Accuracy of Spatial Databases*, London: Taylor & Francis.

João, E., 1991, The role of the user in generalisation within GIS, in Mark, D. and Frank, A. (Eds) *Cognitive and Linguistic Aspects of Geographic Space*, pp. 493–506, Dordrecht: Kluwer Academic Publishers.

João, E., 1994, 'Causes and consequences of map generalisation', unpublished PhD thesis, Birkbeck College, University of London.

João, E., Herbert, G., Rhind, D., Openshaw, S. and Raper, J., 1993, Towards a generalisation machine to minimise generalisation effects within a GIS, in Mather, P. (Ed.) *Geographical Information Handling — Research and Applications*, pp. 63–78, Sussex: John Wiley.

McMaster, R., 1987, The geometric properties of numerical generalisation, *Geographical Analysis*, **19**(4), 330–46.

Müller, J.-C., 1990, The removal of spatial conflicts in line generalisation, *Cartography and Geographic Information Systems*, **17**(2), 141–9.

Müller, J.-C., 1991, Generalization of spatial databases, in Maguire, D., Goodchild, M. and Rhind, D. (Eds) *Geographical Information Systems: Principles and Applications*, pp. 457–75, Harlow: Longman.

Openshaw, S., 1989, Learning to live with errors in spatial databases, in Goodchild, M. and Gopal, S. (Eds) *Accuracy of Spatial Databases*, pp. 263–76, London: Taylor & Francis.

Openshaw, S., Charlton, M. and Carver, S., 1991, Error propagation: a Monte Carlo simulation, in Masser, I. and Blakemore, M. (Eds) *Handling Geographical Information: Methodology and Potential Applications*, pp. 78–101, Harlow: Longman.

Rhind, D., 1991, Data access, charging and copyright, and their implications for GIS, in *Proceedings of the Second European Conference on GIS*, Brussels, Vol. 2, pp. 929–45.

Rhind, D. and Clark, P., 1988, Cartographic data inputs to global databases, in Mounsey, H. and Tomlinson, R. (Eds) *Building Databases for Global Science*, London: Taylor & Francis, pp. 79–104.

Smith, N. and Rhind, D., 1993, Defining the quality of spatial data: a discussion document, in *Proceedings of the Land Information Management and GIS Conference*, University of New South Wales, July, pp. 199–207.

Thapa, K. and Bossler, J., 1992, Accuracy of spatial data used in Geographic Information Systems, *Photogrammetric Engineering and Remote Sensing*, **58**(6), 835–41.

Visvalingam, M. and Whyatt, J., 1990, The Douglas–Peucker algorithm for line simplification: re-evaluation through visualization, *Computer Graphics Forum*, **9**(3), 213–28.

14

The effects of generalization on attribute accuracy in natural resource maps

Marco Painho

The Higher Institute for Statistics and Information Management (ISEGI), New University of Lisbon (UNL), Travessa Estêvão Pinto, Campolide, 1000 Lisboa, Portugal

14.1 Introduction

Map generalization is the simplification of observable spatial variation to allow its representation on a map (Goodchild *et al.*, 1991). Generalization is performed for map display and communication purposes, but also, and perhaps more importantly, for analytical purposes (Müller, 1991). Processes occurring in nature tend to operate at different scales that are far from being obvious. Hence, it becomes important to determine at what resolution a certain process should be sampled.

In vegetation mapping and analysis it is common to define a Minimum Mapping Unit (MMU), intended to establish the level of generalization in interpretation of imagery, i.e. to define the size of the smallest detectable feature for mapping purposes. However, the MMU is rarely applied consistently across map classes or regions. The MMU must be related to a classification system that divides space exhaustively. When the classification system is hierarchical, it is possible to assume that different hierarchies correspond to diverse spatial representations of vegetation, or to different levels of generalization.

The value of a vegetation map lies in its predictive value and, consequently, in the location and attribute accuracy of the objects it represents. It is thus important to assess the quality of a map in terms of its accuracy. Generalization also applies to the simplification of map boundaries during digitizing. Measures of map accuracy do not explicitly consider generalization effects, although excessive line generalization and spatial filtering can greatly reduce the predictive value of a map.

In the following sections, some experiments using vegetation transects are described in order to study the effect of generalization on the attribute accuracy of a small-scale vegetation map. The analysis is geared towards the effects of model-oriented generalization rather than cartographic generalization. The effects of model-oriented generalization have been neglected in much of the literature (Chapter 1) and very few studies have been published concerning the implications on results of using generalized data. The discussion that follows encompasses generalization at the spatial resolution and classification levels.

14.2 GIS: vegetation maps and vegetation properties

Many of the objects that are commonly defined on maps do not have precise sharp boundaries in the real world. Among these, units representing social, demographic, economic, and many other natural and socio-economic phenomena are often represented by *convenient* boundaries. In the case of socio-economic data, arbitrary units like administrative boundaries are regularly used. For natural resource data, such as soils and vegetation, the delineation of polygons is subject to classification definitions and expert judgement.

Vegetation stands are thus subjectively defined units. Vegetation mapping poses several problems. Except for rare situations, pure vegetation stands do not exist and vegetation over space changes more or less in a continuum. Consequently, the delineation of vegetation polygons becomes a subjective process. The accuracy assessment of a vegetation map needs to be made within the framework of a particular classification system that will play the role of a filter through which space is partitioned. The classification of vegetation in order to produce a map will always be subjective, no matter how appropriate it might be for a specific situation, or how careful the scientist might be in applying a certain classification scheme. Many times, even trained botanists, utilizing the same classification scheme, will disagree on the classification and delimitation of a certain stand (Felix and Binney, 1989). Indeed, even if all observers could agree on a single interpretation of a vegetation class defined around a point, they would still divide space into different units both in terms of number and shape. If one adds to the comparison map that was produced using a different classification scheme, then it becomes very difficult, or virtually impossible, to perform any kind of assessment on the positional accuracy of vegetation class boundaries.

In vegetation mapping practice, however, the model of vegetation as a continuum is replaced by a model that, in fact, separates vegetation as a whole into vegetation types assuming, for mapping purposes, that those types can be separated by boundaries. Boundary placement is guided by the class attributes and description as well as, depending on the method used, texture pattern on an image, or field information. Ideally, the resulting polygons should be homogeneous, but in practice they obviously hold different degrees of homegeneity. Given the previous assumptions on how vegetation maps are normally produced, it is natural that many errors occur. Thus, when a certain region is partitioned into vegetation polygons, errors originate because boundaries are placed where they do not exist and polygons are assigned homogeneous classes that are, in fact, heterogeneous.

In order to determine the accuracy of a map such as the one described above, one would need to compare it directly with the truth, i.e. with the vegetation existing on the ground. Unfortunately, such a process is not feasible. A weaker, but still valid alternative, is to compare the vegetation map with other vegetation data considered to be of high accuracy. When these assumptions are made clear, it is then possible to refer to the errors mentioned in the previous paragraph as positional and attribute accuracy, respectively. For a vegetation map, positional accuracy refers to difference between the boundary drawn on that map and the equivalent boundary drawn on a source considered to be of high accuracy. In the same way, attribute accuracy alludes to whether or not class assignments agree or disagree on both the vegetation map and the reference data.

The overall accuracy of a vegetation map depends on many factors such as scale, number of classes, size and shape of polygons. Attribute errors in small-scale vegetation maps can be computed in terms of their overall classification accuracy. The indices provided by such calculations give useful information about map errors in a non-spatial way, i.e. it is possible to understand what particular vegetation classes present more map errors, but the distribution of errors over space is not uncovered. Errors in vegetation maps are expected to vary in space according to a number of factors, including adjacency to particular types, distance to polygon boundaries and terrain type. Furthermore, and considering the subjective nature of vegetation mapping, map errors are predicted to vary with the type of reference data used for the accuracy test: point data, transects or area-class maps. Another error source that can be anticipated in small-scale vegetation maps is extensive generalization that is a direct consequence of the classification system used. Generalization produces more simplified representations and naturally omits polygon heterogeneity, which eventually causes map errors to increase.

Digital vegetation maps will inevitably exhibit some degree of positional error that is introduced during digitizing. While this type of error may be important in GIS applications requiring very precise location information, such as engineering applications, it is expected that in small-scale vegetation maps the magnitude of errors due to digitizing will be small or even cancel out.

The utilization of GIS technology makes it possible to utilize different sets of data, and combine them to produce statistical, as well as graphic reports (Tosta and Marose, 1986). Nevertheless, there are some problems with the accuracy of operations performed by a GIS that stem from uncertainty that is characteristic of most of the data. GIS processes can combine data from different sources with different levels of spatial resolution, using rules that are often very complex. Modelling using such diverse spatial datasets leads, in many cases, to error effects that are difficult to control. The tracking of these errors is a very important issue for the user, who may want to be aware of the origin and extent of the error.

Among all the issues involved in the accuracy of Landuse/Landcover mapping (especially over its vegetation classes), classification is probably the one playing the major role. The problem of classification in vegetation maps can be studied from at least two angles. On one hand comes the choice of a proper classification system for a given area or purpose, which is not straightfoward because classification systems tend to partition a continuously varying entity, vegetation, into discrete well-defined units. In this case, one studies the intrinsic accuracy problems of one classification system as applied to one map. On the other hand, there is the problem of comparing vegetation maps originally produced using different classification systems and/or different scales. In this second case there is a need to study not only each classification system but also how the systems can be cross-checked with minimum accuracy loss.

The digitizing errors that might occur during the input process into the database are again of two types: (1) mislabelling of polygons, when a polygon of class *a* is assigned to class *b*, and (2) mislocation of polygon boundaries due to digitizing error. Due to the nature of vegetation, errors of type (1) are likely to be more important than those of type (2). The boundaries of stands are indeed fuzzy, and in certain cases the aggregate unit area characteristics form part of a continuum (Adejuwon, 1975). It is important to note, that even at large scales, boundaries remain somewhat vague as they are no more than a human artifact representing the real world (Fisher, 1988). Because the boundaries of vegetation are often of imprecise location, the magnitude of the errors due to digitizing is likely to be of lesser importance and to cancel out. However, these errors may become important when the accuracy of a map is tested through the use of field samples that are represented in the GIS by a point (a single x,y coordinate). In this case, it is not inconsequential whether a point lies to the right or the left of a polygon boundary.

14.3 The CALVEG map and its classification system

CALVEG is a hierarchical classification system of actual vegetation combined with descriptive classes, designed to assess vegetation-related resources throughout California (Parker and Matyas, 1981). Table 14.1 illustrates the hierarchical system and its relationship with the descriptive classes.

Table 14.1 Illustration of the hierarchy and the relationship between identified categories and classes in CALVEG

Category	Class
Association (basic unit)	Jeffrey pine/sagebrush/squirrel tail grass
Series	Jeffrey pine, red fir, chamise, sedge-rush
Subformation	Subalpine mixed conifer, montane mixed shrub
Formation class	Conifer forest/woodland. chaparral, herbaceious

The system was devised in the late 1970s by the California region of the US Forest Service to describe and map natural vegetation in the state. A statewide map was produced at the *series* level, with series defined by the dominant overstorey species of the community (e.g. ponderosa pine series). In order to facilitate the classification, the state of California was divided into eight ecological provinces, which contain areas of more or less similar climatic, floristic, and physiognomic factors.

CALVEG mapping was done between 1979 and 1981 by US Forest Service personnel by photo-interpretation of colour infrared prints of Landsat Multi-spectral Scanner (MSS) imagery acquired between 1977 and 1979 (Matyas and Parker, 1980), i.e. image interpretation by existing soil and vegetation maps via 5000 miles of field checking and by personal contact with vegetation experts throughout the state (Matyas and Parker, 1980). The nominal minimum mapping unit was 400–800 acres, but the spatial resolution of the resulting map was very coarse. Average polygon size of the map mosaic (i.e. with boundaries due to map edges deleted) was 38 000 acres for the entire state. The map was field checked at the time of its compilation (Wendy Matyas, 1989, personal communication), but estimates of its accuracy were not published. The map's editors asserted that it was intended only for the broad-scale resource assessments and not for detailed planning (Wendy Matyas, 1989, personal communication).

14.4 Using transect data to assess the effect of generalization on attribute accuracy

14.4.1 Transect analysis of map generalization

In order to assess the effects of generalization on attribute accuracy in CALVEG, a set of transects was generated and overlain on CALVEG. A total of one hundred 4000 m transects were produced. This test site, referred to as the East Bakersfield site, is characterized by a complex mix of conifer forest and woodland of several dominant species, several oak communities, and various shrub and herbaceous vegetation. The site spans the Sierra Nevada from the intensively developed agricultural region of California's San Joaquin Valley to the west, to the Mojave Desert, consisting primarily of creosote bush and Joshua tree communities in the east.

In order to evaluate the accuracy of the CALVEG map, an overall accuracy index (see Table 14.2) was computed. This index normally consists of the ratio of the sum of the diagonal values to total number of cell counts. In this case, the length of the transects was used instead of cell counts. In order to avoid natural boundary variation only the transects not centred on boundaries were used at this stage. They correspond to 200 segments, each 500 m long, grouped into 25 transects of 4 km. The vegetation type along the transects was photo-interpreted, at 500 m intervals, using a classification scheme derived from CALVEG by aggregation into 11 classes (8 of which are vegetation).

Table 14.2 Contingency table comparison of CALVEG (in rows) with vegetation observed along 25 transects of 4 km (in columns). GIS overlay of transect data and the CALVEG map was used to calculate total metres in each combination of map and ground class

	Photo transect									
CALVEG	CF	BF	CW	BW	Sh	DS	Hb	Ag	Total	% Agreement
CF	16219	500	4500	4170	2000	0	500	0	27890	58.2
BF	0	0	1552	0	111	0	1500	122	3285	0
CW	12430	0	5455	501	6934	0	1000	0	26320	20.7
BW	1500	0	948	5199	1889	0	3846	378	13759	37.8
SW	0	0	0	0	0	0	3417	0	3417	0
Sh	354	0	1000	2237	9500	0	4235	0	17326	54.8
Hb	0	0	0	2502	0	1495	3500	16	7512	46.6
Ag	0	0	0	0	0	0	0	484	484	100
Total	30503	500	13455	14609	20434	1495	17998	1000	99994	

Correct = 40357/99994; Correct = 40.36 per cent; expected correct = 18.97 per cent; *kappa* = 26.39 per cent.

Key: CF, conifer forrest; SW, succulent woodland; BF, broadleaf forest; Sh, shrubland; Wa, water; CW, conifer woodland; DS, dwarf shrub; Ag, agriculture/urban; BW, broadleaf woodland; Hb, herbaceous.

The total percentage correct (or agreement) is 40 per cent and *kappa*[1] (Rosenfield and Fitzpatrick-Lins, 1986) is 26.39 per cent. These values are very low considering the large spatial scale and the simple vegetation classification scheme. The low values are mainly due to disagreements in types: broadleaf woodland, conifer woodland, broadleaf woodland and herbaceous. Such a disagreement can probably be explained by the fact that 500 m transects depict land-cover detail that is not present at the scale that CALVEG was mapped. In other words, the 500 m interval could still be below the MMU of CALVEG.

In order to test this hypothesis, four moving windows of sizes 500, 1500, 2500 and 4000 m were passed along the transects and the percentages correct and incorrect were computed. The transects not centred on conifer boundaries were divided into two sets: (1) transects contained inside conifer polygons as defined in CALVEG (14 transects), and (2) transects contained in non-conifer polygons (11 transects). For conifer polygons, the results show (see Figure 14.1) a slight increase in agreement with longer windows, maybe suggesting that smaller windows depict detail not mapped in CALVEG. However, and contrary to what was expected, in non-conifer polygons (see Figure 14.2) the behaviour seems to be opposite: there is a slight decrease in agreement with increased window size, showing that map errors for these types in this region are mainly due to polygon misclassification and not to generalization.

In an attempt to further understand classification error in the CALVEG map, the transects located in conifer forest and conifer woodland polygons were aggregated in 500, 1000 and 2000 m blocks. Figures 14.3 and 14.4 depict the distribution of vegetation types along the transects contained in conifer forest and conifer woodland, respectively. Both conifer forest and conifer woodland polygons can be expected to reveal a spatial distribution across very large areas that are not completely homogeneous. The errors referred to in Figures 14.3 and 14.4 may just represent the natural occurrence of other interspersed vegetation types that are not large enough to be represented in a map at the scale of CALVEG.

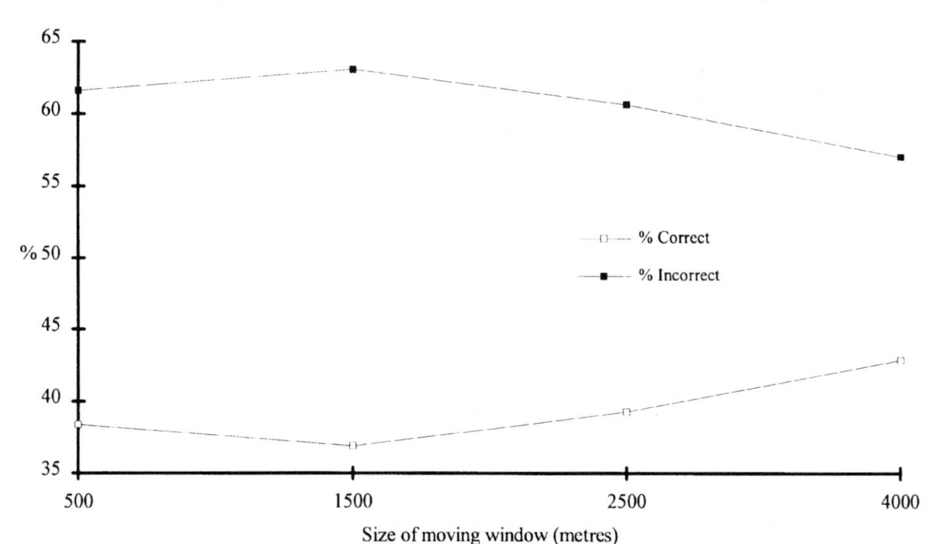

Figure 14.1 Percentage agreement between areas mapped in CALVEG as conifer-dominated vegetation and 1987 vegetation cover photo-interpreted along fourteen 4 km transects as a function of the length of transect used for the comparison.

[1] The *kappa* statistic is computed according to Rosenfield and Fitzpatrick-Lins, 1986. The *kappa* statistic is a coefficient of agreement that removes chance agreement before overall percentage correct is computed. Hence the *kappa* statistic is normally smaller than the total percentage correct.

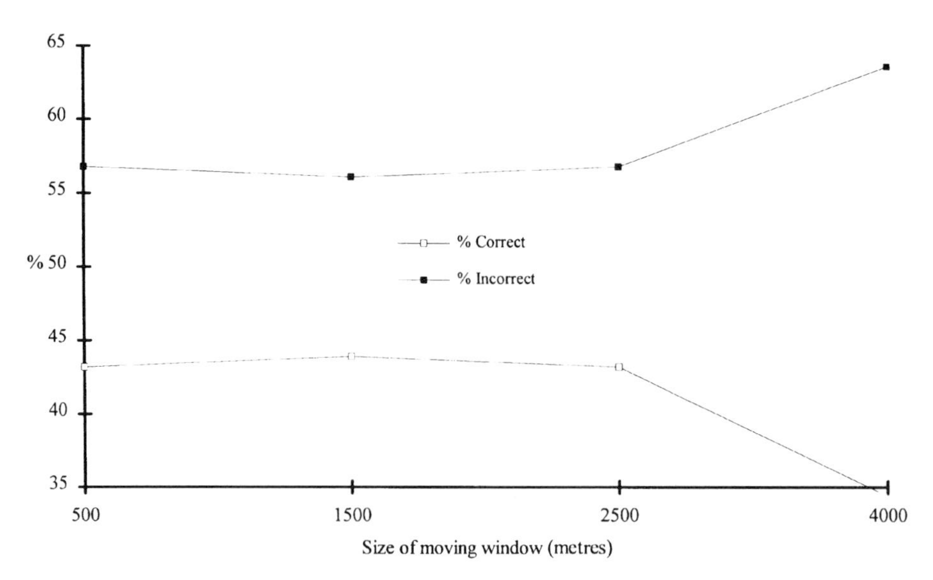

Figure 14.2 *Percentage agreement between areas mapped in CALVEG as non-conifer vegetation and 1987 vegetation cover photo-interpreted along fourteen 4 km transects as a function of the length of transect used for the comparison.*

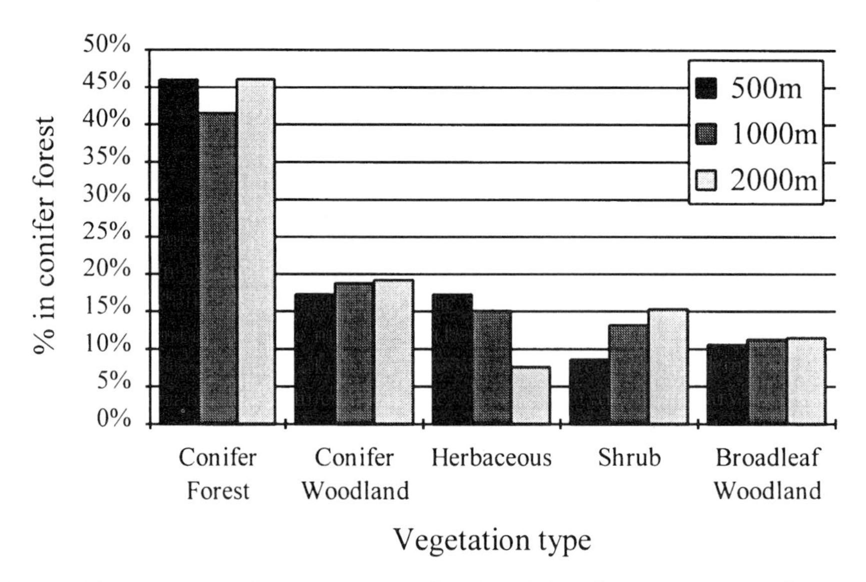

Figure 14.3 *Frequency of vegetation types for 500, 1000 and 2000 m segments for transects located in CALVEG conifer forest.*

14.4.2 Sampling frequency and map generalization

As have been discussed before, fundamental properties of vegetation maps include their classification scheme and their spatial level of detail or scale. One measure of scale is given as the smallest feature that is represented on a given map. To determine the sensitivity of a small-scale vegetation map to Minimum Mapping Unit (MMU), vegetation transects from several regions of southern California were analysed. The effects of

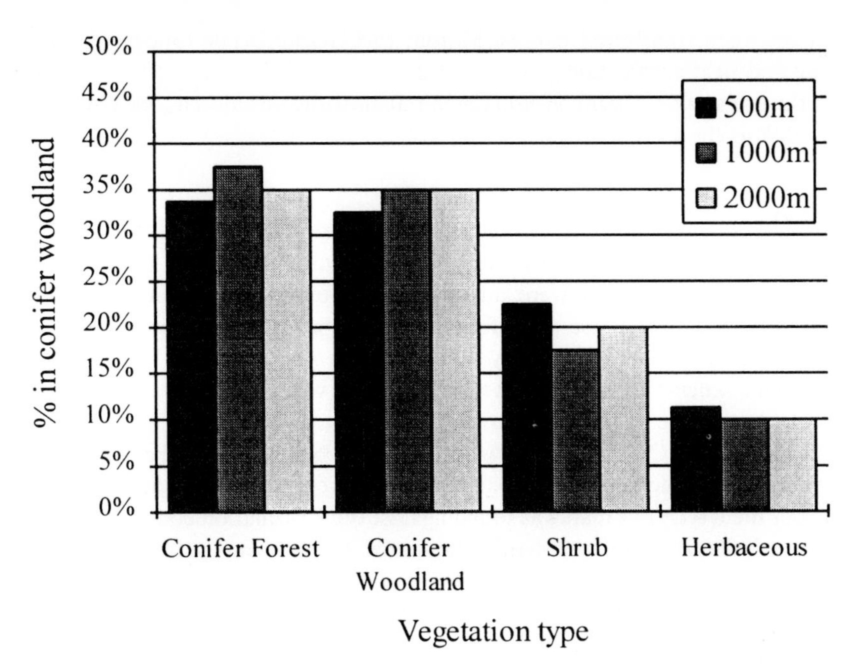

Figure 14.4 Frequency of vegetation types for 500, 1000 and 2000 m segments for transects located in CALVEG conifer woodland.

MMU on map classification could then be tested by representing transect vegetation at different block sizes. The determination of the largest acceptable MMU has obvious practical significance.

Data gathering

A stratified random set of points were selected within southern California such that one point was located within each 15 minutes of latitude and longitude. The Pascal code which generated the point coordinates was written so the latitude and longitude of each random point could be directly input to ARC/INFO. Working plots were then printed at a scale of 1:250 000 for overlay on USGS quadrangle maps. Each of the points from the paper plots was used to identify the closest photo-centre location from the National High Altitude Photography Series (NHAP) 1:58 000 scale air photos. Transects or transect sections that occurred in urban environments were omitted from the study. The displacement of points to the photo centres allowed the maintenance of one photo per point and additionally means transects are in the least distorted portions of each photo. On each photo centre, a linear 8 km transect was centred consisting of 64 segments of 125 m. Due to variations in topography, elevation, and aircraft altitude, the mean transect segment was expected to be different than the calculated 125 m. Transect direction was alternated between north–south and east–west with each photo to minimize topographic bias. Using stereo-photo pairs from the positive transparency air photos, each 125 m section was manually photo-interpreted using the CALVEG, Kuchler (Kuchler, 1977) and WHR (Wildlife–Habitat Relationships: Mayer and Laudenslayer, 1988) classification schemes. The WHR classification system divides the state into five formations and 48 types. The transect was then transferred from the photo to USGS 1:24000 topographic maps for digitizing and field verification.

Since the transect data was coded in the WHR classification system, the CALVEG map was recoded to the same system.

Data analysis

In the previous section, transects were used to determine overall map accuracy for CALVEG. The classes used were an aggregation of CALVEG into 11 types, 8 of which were vegetation. These were further aggregated in two types only in order to test the variation of overall accuracy with sampling size (at 500 m intervals). Using a similar binary scheme (conifer forest versus conifer woodland) it was also possible to analyse the homogeneity of these two broad types.

In this section, the CALVEG map will still be used but now it will be compared with a dataset that is more detailed (125 m) and can be aggregated up to 4000 m. The attribute accuracy of the CALVEG map was tested against the information coded in the transects. Two types of generalization were tried. One spatial, by reducing the transect resolution to intervals of 250, 500, 1000, 2000 and 4000 m, and the other by aggregating WHR types to which CALVEG had been previously converted.

The contingency table analysis of transects versus CALVEG reveals a quite low level of accuracy. For the resolution level of 125 m, the agreement is only 21 per cent and *kappa* 17 per cent. A summary of the resulting confusion matrix is presented in Table 14.3. Overall, and considering the classes aggregated in broad types, such as shown in Table 14.3, conifer types tend to be mapped more accurately (69.8 per cent agreement) when compared to hardwood (35.6 per cent) and shrubs (49.6 per cent). The major confusion is registered among hardwood types and all the other types together. Many of the errors, such as those confusing hardwood woodland types with herbaceous, and montane hardwood–conifer (MHC) with sierran mixed conifer (SMC), may be due to polygon misidentification, since these types are often hard to distinguish. Other errors are due to map generalization since they appear between types that occur in areas of similar ecological

Table 14.3 Contingency table of CALVEG (columns) and transects (rows) at the maximum level of generalization (125 m)

	CALVEG						
Transects	Conifer types	Hardwood types	Shrub types	Herbaceous types	Other	Total (m)	Agreement (%)
Conifer types	**48915**	1505	8885	0	10725	70030	69.8
Hardwood types	21130	**34505**	23390	7875	10000	96900	35.6
Shrub types	18005	13770	**61924**	4520	26740	124959	49.6
Herbaceous types	2125	16380	6140	**10900**	12050	47595	22.9
Other	3380	4505	19005	2875	**23305**	53070	43.9
Total (m)	93555	70665	119344	26170	82820	**392554**	
Agreement (%)	52.3	48.8	51.9	41.7	28.1		

Table 14.4 Levels for kappa *and total percentage correct for different levels of transect generalization*

Length (m)	% Correct	*kappa* (%)
125	20.5	16.9
500	20.5	17.2
1000	22.3	18.1
2000	20.6	16.0
4000	23.3	21.5

conditions. Examples of these may be sierran mixed conifer (SMC) and ponderosa pine (PPN), or chamise-redshank chaparral (CRC) and mixed chaparral (MCH).

In order to test the effects of spatial generalization on the attribute accuracy, the transects were aggregated to decreasing levels of spatial detail: 250, 500, 1000, 2000 and 4000 m. The results show that the classification accuracy was not sensitive to this type of generalization (Table 14.4). Both the total percentage correct and *kappa* remain more or less stable throughout the levels.

An aggregation at the class level was then tried. According to the CALVEG hierarchical scheme, the 34 initial WHR types were aggregated into 11 types: conifer forest/woodland, hardwood forest/woodland, chaparral, soft chaparral, sagebrush shrub, desert shrub, dwarf scrub, herbaceous, and three types of non-vegetative types — barren, water, and urban–agricultural. Contingency tables were again computed for all levels of possible transect lengths. There was a notable increase in the total percentage of agreement (from 21 per cent with 34 classes up to 38 per cent), as well as in *kappa* (from 17 to 27 per cent). This increase in accuracy was expected due to the aggregation of the vegetation classes. Again, in this situation, the variation of *kappa* and total percentage correct did not vary noticeably when CALVEG was compared with the transects at different resolutions, although the variation is more consistent with the 11 types than with the 34 types, indicating again that errors are due to polygon misidentification rather than to generalization. Figure 14.5 shows the behaviour of *kappa* in both experiments.

In order to understand better the effects of spatial generalization, the transects were again used to count the total number of vegetation classes present at different resolutions. The results (Figure 14.6) show a dramatic decrease in the total number of classes, from a maximum of 34 at 125 m resolution down to 23 classes at 4000 m. The decrease is particularly important at transect lengths greater than 500 m.

These results demonstrate how the predictive value of vegetation maps changes with the resolution (equivalent to the Minimum Mapping Unit) at which the stands are mapped. Also the length (a surrogate for area) predicted for each individual class varies according to the resolution at which it is measured. The change in total transect length is not consistent for all classes. For example, annual grassland (AGS) shows a decrease of 13 per cent in length from 125 m (44 750 m) to 1000 m (39 000 m) resolution, whereas montane hardwood (MHW) increases 53 per cent from 250 m (23 500 m) to 4000 m (35 935 m) resolution. Overall, the 50 transects cover a large spatial extent that encompasses regions of very different ecological and climatic conditions.

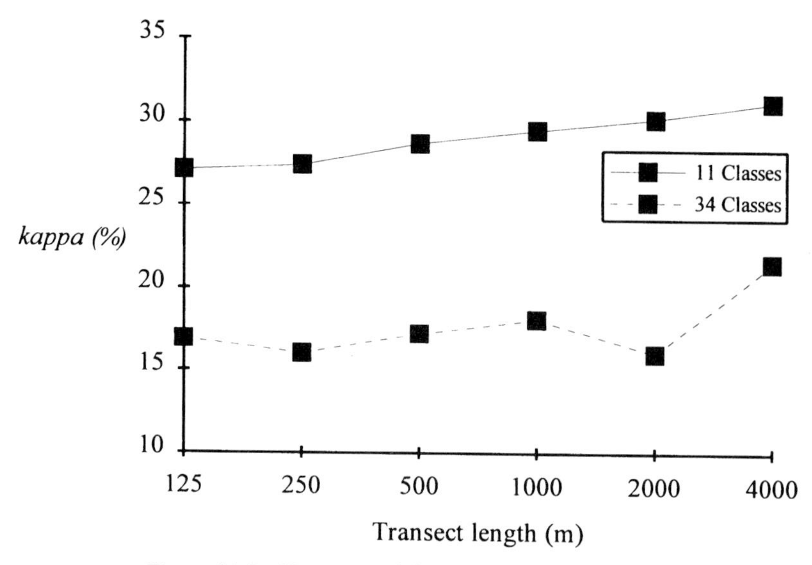

Figure 14.5 Variation of kappa *with transect length.*

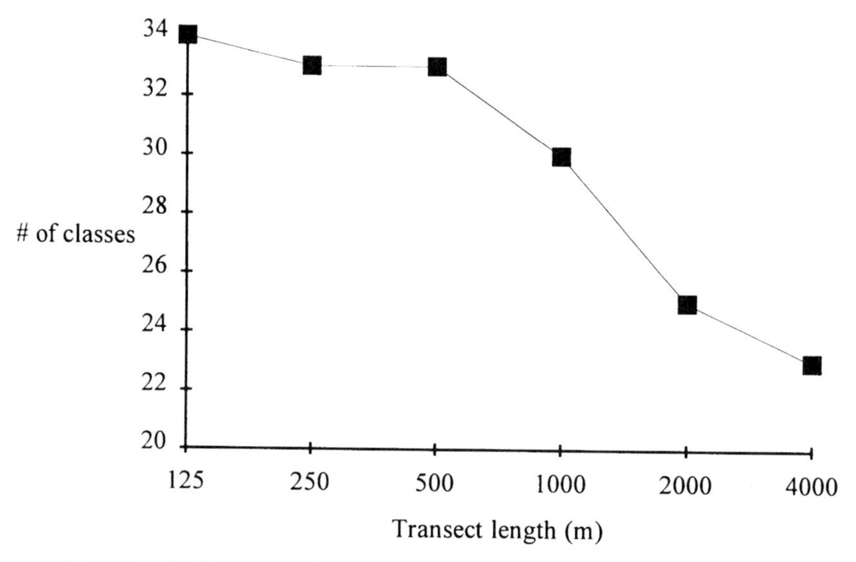

Figure 14.6 Variation of total number of WHR classes with transect length.

14.5 Summary

Attribute errors in CALVEG polygons can be attributed to many different reasons. Some of those include natural human error in digitizing labels and lines, as well as errors in identifying correct types during photo-interpretation. However, even if it is assumed that such mistakes do not exist, CALVEG would always present attribute errors when compared to datasets that exhibit more spatial resolution or a classification system incorporating more types, especially if such a classification is not hierarchically compatible.

Class generalization of the test data, by hierarchical aggregation, generally had the effect of increasing attribute accuracy on the CALVEG map. However, generalization at the spatial level (allowing transects with increasing lengths) did not consistently affect accuracy of the tested map to a great degree, suggesting that the error source is polygon mislabelling rather than generalization. Different classes respond in different ways to generalization; in some the accuracy increases, in others it decreases. It is thus not possible to generalize about the effect of generalization on accuracy as a whole, at least at the scales of measurement that were used here. However, the situation might change if larger datasets with larger spatial extent were used.

An important effect of generalization was the change in the predictions of how many types exist in a region and how much area they occupy. It was shown that the number of classes rapidly decreases as the minimum mapping unit (or transect length) increases. Another important factor is, of course, the vegetation model that was used throughout the analysis. In the vector model, even when measuring accuracy, it is assumed that vegetation is homogeneous over space within the polygon boundaries. Tests with the first set of transects (East Bakersfield) showed how heterogeneous the vegetation polygons are, often incorporating many other closely related types and even types belonging to other formations.

The use of generalized data in modelling natural phenomena has implications that are far from being clearly understood. Generalization in vegetation mapping and analysis needs further research especially in determining the relationships between generalization, scale and accuracy. The results of the experiments carried out with the CALVEG map show that even some accepted statements may be wrong. One such case would be that analysis of a map against test data with increasing class generalization should increase the map's class accuracy. The experiments showed that this is not always the case.

If one is looking for scaleless geographic databases one must carefully define how changes in scale (generalization) really affect data, especially in systems that are heavily dependent on a classification system for representation, such as vegetation soils or geology. Furthermore, a change in scale for different variables must necessarily follow different rules.

The massive dependency on analogue maps for data gathering and referencing is probably an important error source and the major obstacle to model-oriented generalization. The substitution of traditional sources by Remote Sensing, Global Positioning Systems, and others to come, will probably help to solve many of our current problems.

Acknowledgements

The work presented in this paper was partially supported by the following institutions: JNICT, Junta Nacional de Investigação Científica e Tecnológica, Portugal; SEARN, Secretaria de Estado do Ambiente e Recursos Naturais, Portugal; CDFG, California Department of Fish and Game; CDF, California Department of Forestry; ESRI, Environmental Systems Research Institute, Redlands, California; NCGIA, National Center for Geographic Information and Analysis.

References

Adejuwon, O., 1975, A note on the comparison of chorochromatic surfaces, *Geog. Anal.*, **7**(4), 435–40.

Felix, N.A. and Binney, D.L., 1989, Accuracy assessment of a LANDSAT-assisted vegetation map of the coastal plain of the Arctic National Wildlife Refuge, *Photogramm. Eng. Remote Sensing*, **55**(4), 475–8.

Fisher, P.F., 1988, 'Knowledged-based approaches to determining and correcting areas of unreliability in geographic databases', presented at the First Specialist Meeting of Research Initiative One of the National Center for Geographic Information and Analysis, 12–16 December.

Goodchild, M.F., Davis, F.W., Painho, M.O. and Stoms, D.M., 1991, The use of vegetation maps and Geographic Information System for assessing conifer lands in California, *National Center for Geographic Information and Analysis/NCGIA Technical Paper 91–23*.

Kuchler, A.W., 1977, A map of the natural vegetation of California, in *Terrestrial Vegetation of California*, 1st Edn, Barbour, J. and Major, J., (Eds), pp. 909–38, New York: John Wiley.

Matyas, W.J. and Parker, I., 1980, CALVEG: mosaic of existing vegetation of California, San Francisco, CA: Regional Ecology Group, US Forest Service.

Mayer, K.E. and Laudenslayer, W.F. (Eds), 1988, *A Guide to Wildlife Habitats of California*, p. 166, Sacramento, CA: California Department of Forestry and Fire Protection.

Müller, J.-C., 1991, Generalization of spatial databases, in Maguire, D.J., Goodchild, M.F. and Rhind, D.W. (Eds) *Geographical Information Systems*, Essex: Longman.

Parker, I. and Matyas, W.J., 1981, CALVEG: a classification of California vegetation, San Francisco, CA: Regional Ecology Group, US Forest Service.

Rosenfield, G.H. and Fitzpatrick-Lins, K., 1986, A coefficient of agreement as a measure of thematic classification accuracy, *Photogramm. Eng. Remote Sensing*, **52**(2), 223–7.

Tosta, N. and Morose, R. 1986, The distribution of California hardwoods: results of a statewide geographic information system, in Plumb, T.R. and Pillsbury, N.H. (Eds) *Proceedings of the Symposium on Multiple-Use Management of California's Hardwood Resources*, San Luis Obispo, CA: Pacific Southwest Forest and Range Experiment Station, pp. 304–8.

SECTION VI

Operational Issues

15

Incremental generalization for multiple representations of geographical objects

Tiina Kilpeläinen and Tapani Sarjakoski

Department of Cartography and Geoinformatics, Finnish Geodetic Institute, Ilmalankatu 1 A, SF-00240 Helsinki, Finland

15.1 Introduction

Geographical information is collected and updated at the base level of a geodatabase. The information needed is gathered from the geographical databases for different applications. However, the demands of different data do not presume that all the data should be saved in the databases. The data not available in the database must be transformed and produced from the existing one. At the same time, multiple views and representations of the same data should be possible. The challenge will be to try and find a flexible data model in which to store the data. This data model should be able to satisfy even the demand of multiple representations of the same data (Kilpeläinen, 1992). The central concern is how to make updated representations from the base level to the multiple abstraction levels. It is evident that methods for fully automatic generalization do not yet exist, neither are there methods that would maintain geographical databases at multiple representation levels in real time. However, one is forced to face the practical problems on everyday use of geographical data, as, for example, the necessity to make updates to the base level of multiple representations in the geodatabase.

This work studies under which conditions the updates on the base level of the geodatabase could be propagated to the other geometrical abstraction levels. The task of creating more abstract representations starting from the base level constitutes a model generalization process. This propagation of updates is called here incremental generalization. Automatic incremental generalization would make the whole generalization process automatic in a multiple representation environment. The idea assumes that the generalization problem is divided into modules. The question that arises here is whether this is possible?

Up to now the generalization problems have mostly been concerned with the cartographic visualization problems, e.g. how to visualize the amount of information that has been visualized on a target scale, instead of on a smaller scale, so that we would simplify the layout of the information but keep the characteristics of the information, in order that the cartographic message would remain the same. Mark (1991) states that the importance of generalization is not only for graphic display but for efficient and appropriate spatial analysis, too. This epoch of geodatabases demands that the role of the visual representation be evaluated again. The generalization problem should also be examined separately from the visualization context, as a conceptual problem. As the conceptual nature of the generalization process is emphasized, the main problems of generalization are related to the understanding process.

In the first part of this chapter the multiple representation problem is briefly discussed. Then the authors' idea of incremental generalization is introduced as an alternative approach for updating problems in multiple representation geodatabases. The term incremental generalization has its background in software engineering, which is discussed as well. Finally, the problem of how to divide generalization problems into modules is examined.

15.2 Multiple representation problem and generalization

15.2.1 Generalization as a conceptual problem

Cartographic generalization means the variety of modifications that can, and must, be made as a result of the reduction of information (Robinson *et al.*, 1978). At the same time the effectiveness of cartographic communication is increased. In Kilpeläinen and Sarjakoski (1993) it is indicated that generalization should be regarded as a part of the cartographic communication process. The generalized output is always specific for the application. The conceptual basis for the generalization process should originate from modelling of the real world entities. It is essential that the cartographic message that should be communicated to the user is modified in the conceptual generalization phase, and should not be modified in the visualization phase (Kilpeläinen, 1992). What is meant, for example, by aggregation of buildings? The geometric interpretation is that one is moving from point representation into areal representation. This is then visualized by certain graphical representations, for example, with grey area. The conceptual generalization that has taken place is related to the changes in geographical meaning — from buildings into blocks. The generalization problems should not be disregarded in connection with visualization, but a greater effort should be made to understand the generalization problem separately from the visualization context, as a problem of spatial changes varying from one representation resolution to another.

15.2.2 Multiple representation in the database environment

Multiple representation is closely connected with generalization problems in the database environment. The geographical database must include representations of geographical objects at different levels. One reason for multiple representations, given by Jones (1991), is the relatively limited capabilities for automatic generalization. Besides, the geographical database should support modification across the different resolution levels so that changes applied to one level can propagate to the others. The requirements for allowing other resolution levels to be deduced automatically, according to Bruegger and Frank (1989), is that the connections between levels are formally described, as well as the objects and their relationships at all levels.

Multiple representation is in no way a new problem. It has also been recognized earlier, in practice, in several other applications. For example, in the Finnish town Espoo, in connection with modelling the data management system for the city, the problem was identified (Espoo City, 1988). Nevertheless, realizing the strong relations to the generalization problem in the database environment especially, has recently made the multiple representation problem an actual research topic (Buttenfield and Delotto, 1989; NCGIA, 1989). The challenge one faces is to describe spatial knowledge so precisely that it can be processed by a computer. The level of formalization must offer the possibilities for deducing new information from the existing one.

The question today is not so much if these multiple representations are necessary, but rather how this can be done. In the following, an alternative proposal is made in order to attack the updating problems in the database environment: incremental generalization.

15.3 Incremental generalization

15.3.1 The principle — background in software engineering

The principle of incremental generalization relies to some extent on the methods used in software engineering and program development. One of the goals in software engineering has been to divide the task of program development into subtasks. This is called the modular approach.

Different epochs in program development may be recognized. One of them can be called the Fortran approach. The program was developed so that it consisted of subprograms. These could be developed and compiled independently, and then linked to form an executable program.

This approach suffered from the fact that it did not assure the program to be internally consistent, even at the syntax level. This problem was then solved with the development of the Pascal language. In the original Pascal language, a program was considered to be a whole. The whole program had to be recompiled even if only a single line was changed. This approach definitely guarantees the syntactical consistency of the program but at the same time it increases the compilation time, thus making it impractical.

The benefits of these two approaches are combined in programming languages like Ada and Modula 2 (Ledgard, 1983). The central principle in them is that each module is strictly divided into two parts: module interface and module body. This allows the compiler environment to make the compilation incrementally, which means that only the modules that have been modified, or which are dependent on the modified ones, have to be recompiled. Dependency information is maintained automatically in the program development environment to support the consistency checks. As a result, the program can be compiled incrementally in modules and at the same time its consistency is guaranteed.

In this chapter, the authors wanted to study how the principle of incremental compilation could be applied analogously in a generalization task.

15.3.2 Generalization as a batch process

The notation that is used in the following figures is adopted from the so-called SA/SD method (DeMarco, 1979; Ward and Mellor, 1985 and 1986); see Figure 15.1. Figure 15.2 shows the scope of update transactions.

SA/SD Method

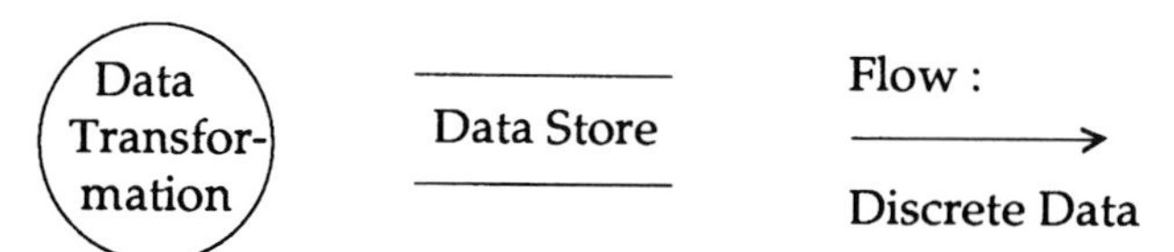

Figure 15.1 Notation according to the SA/SD method.

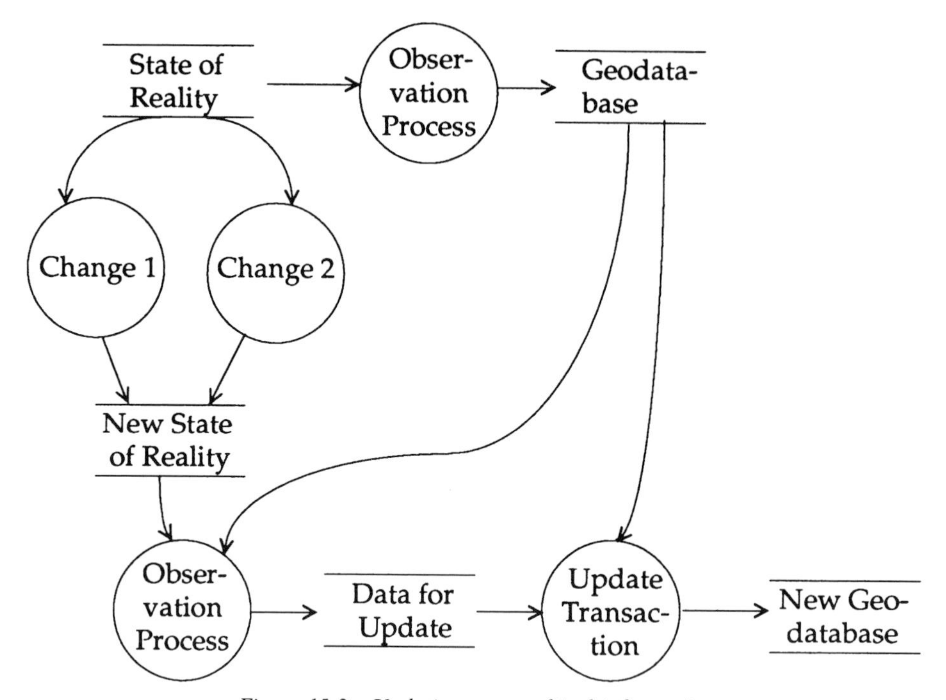

Figure 15.2 Updating geographical information.

Beginning with state of reality observations on a certain state of geographical reality stored in geodatabase; then, in practice, update transactions are needed. New buildings and roads are built, for instance. Changes in the geographical reality result in a new state of reality. This new state of reality has to be mapped by observation processes in the field or by photo-grammetric data capturing and we get data for updates. These updates require transactions into the geodatabase.

The update transaction process results in a new geodatabase. Different applications may require different kinds of generalization processes that have to be carried out as a totally renewed process. Therefore, in batch generalization, in order to get the different kinds of generalized representations, all the data from the updated geodatabase have to be completely processed again. The central problem faced here is the various versions of generalized representations: those which are deduced from the original geodatabase and those which are deduced from the geodatabases that are updated at different times from different states of reality. Figure 15.3 illustrates how batch generalization is applied for different states of the geodatabase.

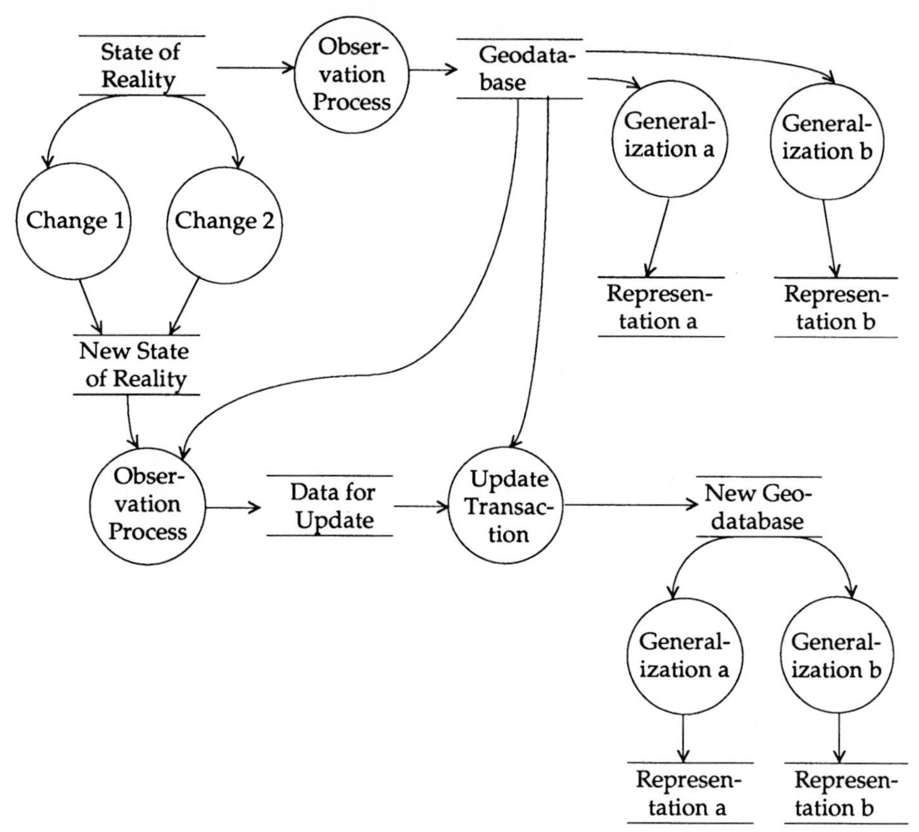

Figure 15.3 Different representations produced by renewed generalization in batch generalization.

15.3.3 Updating in an incremental environment

As is mentioned earlier, in practice, it is necessary to make updates to the base level of multiple representations in the geodatabase. This work studies under which conditions these updates could be propagated to the other levels. In this work, this propagation of updates is called incremental generalization. Here, the authors try to apply the principle of incremental compilation from software engineering analogously in a generalization task.

Automatic incremental generalization would make the whole generalization process automatic in a multiple representation environment. Considering the principle of incremental generalization the question is whether the batch generalization process illustrated in Figure 15.3 could be solved in an alternative way? Figure 15.4 shows the principle of incremental generalization.

In an incremental generalization environment, the generalization process is performed completely for the whole geodatabase only once. After following update transactions to the geodatabase, the old generalized output is also updated in an incremental way, which means that the generalization process is performed only for the modules influenced by the updates. Incremental generalization thus requires that the problem can be divided into modules. This demands that it is clearly defined as to which object classes the changes do not influence at all. Aasgaard (1992) defines a *conflict set* as a set of conflicting objects that may affect the visibility or interpretability of the current object. In a multiple

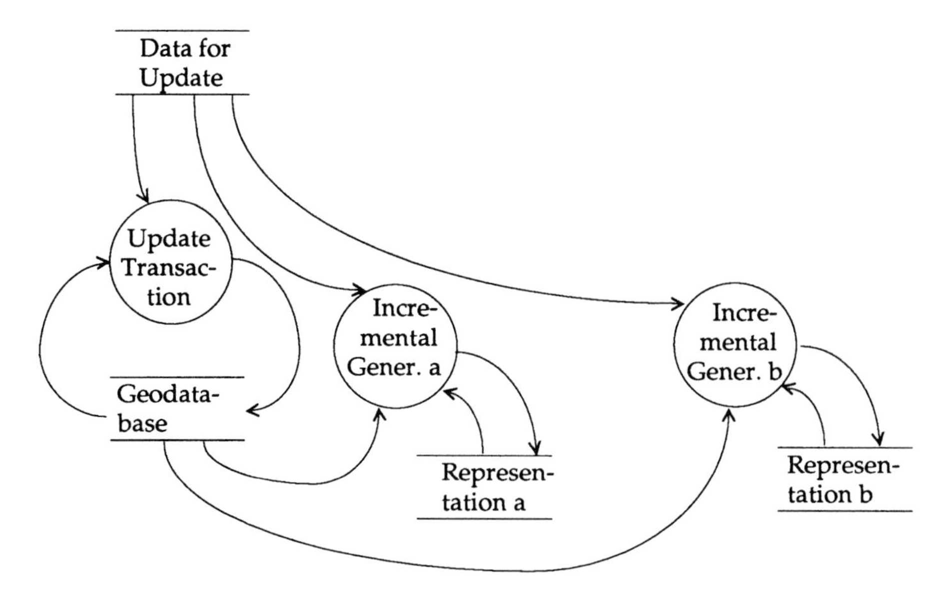

Figure 15.4 The principle of incremental generalization.

representation database, it should be defined for each object which conflict set it belongs to, e.g. the relations of the objects influencing each other. This is the basis for an incremental way of working. In the following, the problem of how to divide the generalization problem into modules is discussed.

15.4 Dependencies in generalization tasks

15.4.1 Other related works

It is an interesting question whether the problem of generalization can be broken up into modules so that it is formally fully defined which geographical object classes influence each other. A certain theoretical basis can be found in the literature for how the objects and their relationships are influenced through the generalization process.

McMaster and Shea (1992) discussed spatial and attribute transformation characteristics. In Shea and McMaster (1989) the application of how to generalize was focused on twelve types of generalization operators:

simplification	merging	enhancement
smoothing	collapse	displacement
aggregation	refinement	classification
amalgamation	exaggeration	symbolization

These include the ten spatial transformations, and the two attribute transformations: classification and symbolization. A list of dependencies on other operators is given by McMaster and Shea (1992, p. 67). Merging and refinement are not dependent on other operators while the rest of the operators are dependent.

Aasgaard (1992) presents a conflict set theory according to which the generalization operations may be classified as possibly conflict generating or non-conflict generating. Refinement, simplification, aggregation, amalgamation, and collapse, will usually not increase the number of conflicts, while exaggeration, and displacement, most certainly will. In Aasgaard's work the conflict set consists of objects that may affect the visibility or interpretability of the current object.

In the literature some proposals for fixed generalization orders are found. For example, water areas should be generalized first, then traffic connections and lastly buildings. In those national agencies where generalization processes are automatically run, this kind of ordering is often applied to a large extent. Fixing the order for a generalization process reflects that it is already quite commonly known how the contents of graphical map elements in map production influence each other.

15.4.2 Ordering generalization steps

In incremental generalization, the set of objects that are not influenced by the update transaction process should be notified in the multiple representation database. The first step in determining this set is to fix the order of generalization steps. This order might be different for different kinds of applications. The problem is two-sided: the generalization process can be attacked not only from the viewpoint of the order of the objects that have to be generalized in a fixed order, but also from the point of view of the generalization process execution (Brassel and Weibel, 1988). If we first look at the order of the objects to be generalized, the order puts the objects into a hierarchy that defines which of the objects influence each other, and which do not interact. This is demonstrated in the following section. Second, the conflict sets for objects stored in the geodatabase have to be identified. As is mentioned above, the generalization operators also influence each other. Now, it has to be analysed what requirements a fixed generalization order puts on generalization operators, e.g. which of these are needed.

15.4.3 Generalization in modules

The principle is that in order to make the incremental generalization operational, it is first necessary to divide the generalization problem into modules. In the following generalization processes, only the modules that are influenced according to the definitions in conflict sets have to be processed. It is not even necessary to make the global generalization task fully automatic, if it is possible to make automatically the changes due to the updating transactions of the geodatabase. This idea is demonstrated in Figures 15.5 and 15.6, in a simplified way, which also demonstrates the importance of object hierarchy.

The following generalization hierarchy is used for the generalization task in this simple example:

1. water areas
2. road network
3. buildings

In Figure 15.5 buildings are updated in the geodatabase. Because the other objects are above the buildings in the hierarchy, these were not influenced and only the buildings were generalized incrementally.

In Figure 15.6 update transactions on roads and buildings are made into the geodatabase. In this case, roads lie above buildings, but water areas are beneath them in the hierarchy. Therefore, the generalization is processed in two modules: roads and buildings. In this example, the generalization operators aggregation and refinement are used. Of course, other operators could also be used, if the application sets demand for it.

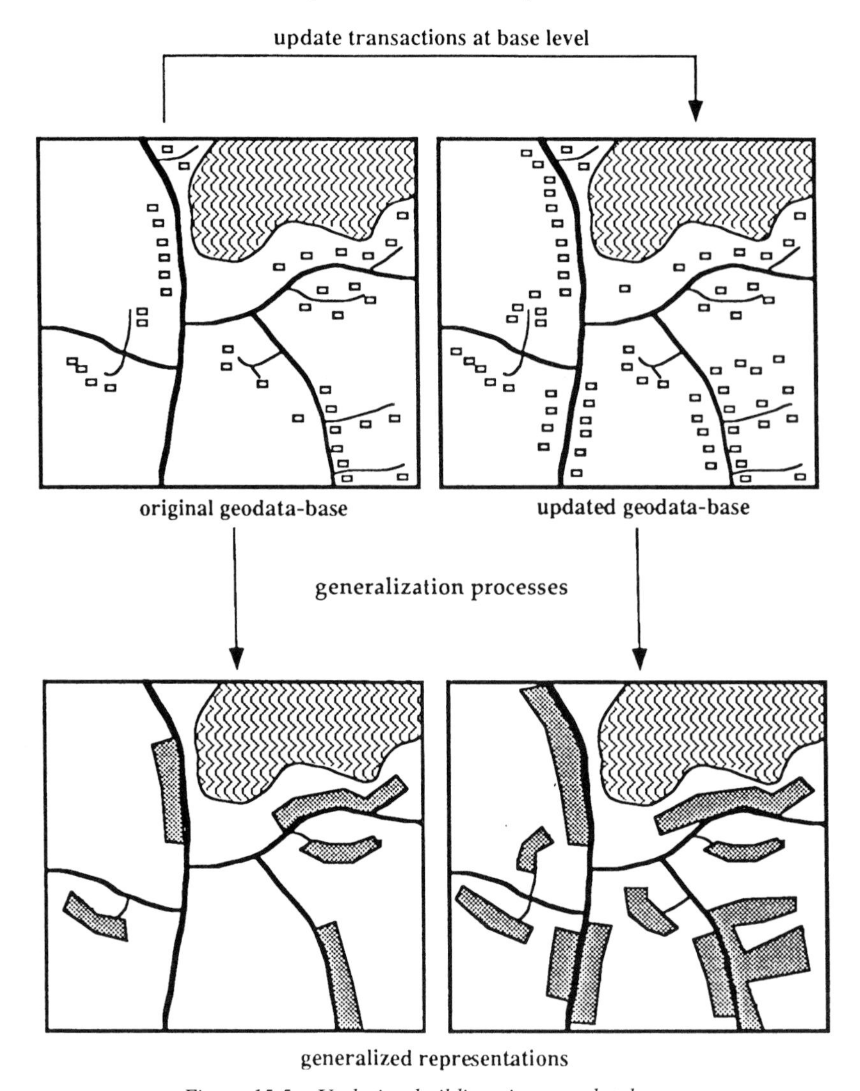

Figure 15.5 Updating buildings in a geodatabase.

The difficult problems encountered are how to define the set of conflicting objects, and how to find the border for the module to be processed? It is evident that the dependency between objects cannot be managed with the simple hierarchical way described above, but that a higher level of structural intelligence is needed.

15.5 Summary

In this chapter, the authors proposed how the principle of incremental compilation could be applied in a generalization task. The principle of incremental generalization relies on the methods used in software engineering. In modular approach dependency, information is maintained automatically. Automatic incremental generalization would make the whole generalization process automatic in a multiple representation environment. The

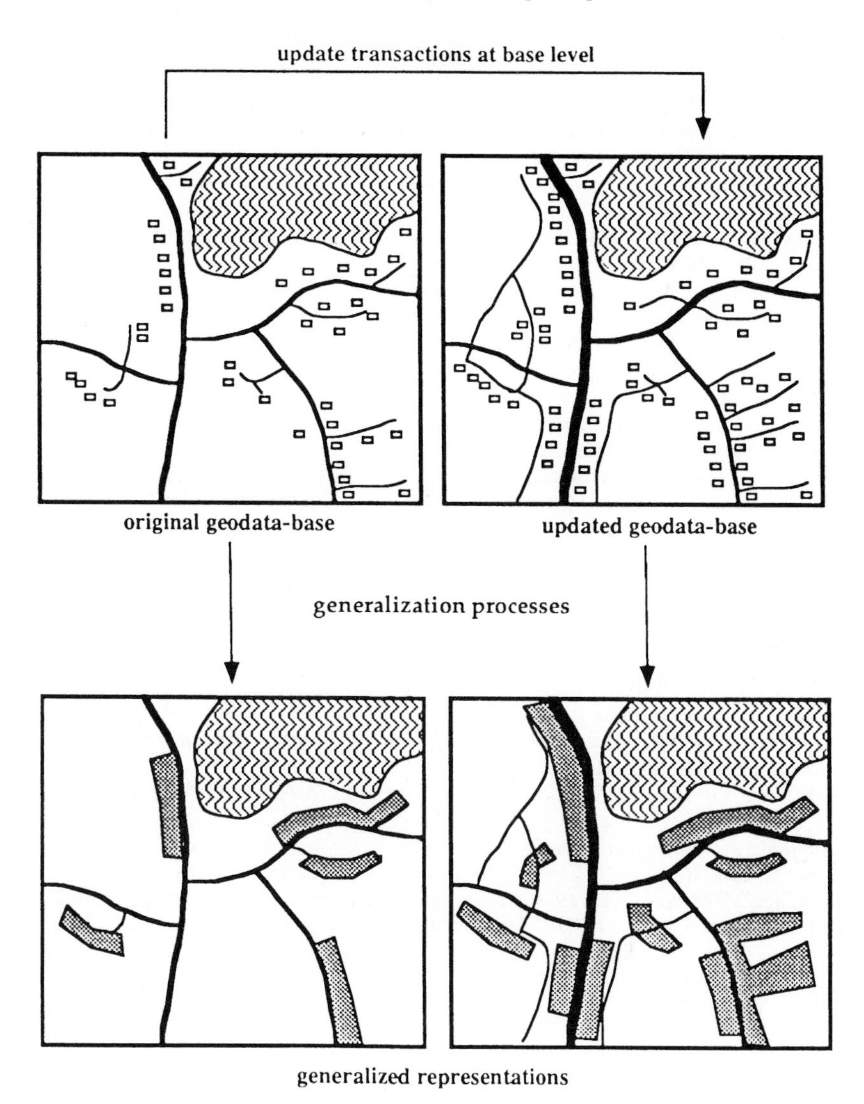

Figure 15.6 Updating buildings and roads in the geodatabase.

benefits of incremental generalization lie also in flexibility. The global generalization task does not even have to be made fully automatic if it is possible to make the changes due to the updating transactions of the geodatabase automatically.

One of the requirements for an operational incremental generalization is to divide the generalization problem into modules. The first step is then to find a generalization hierarchy. Analysis is also needed on dependencies of generalization operators. Incremental generalization can be realized only if it is possible to define the set of conflicting objects. The relationships of the objects influencing each other should be defined in the multiple representation database. Conceptualizing the generalization problem would help to reduce the problem of conflicting objects.

References

Aasgaard, R., 1992, 'Automated cartographic generalization, with emphasis on real-time applications', Doctor's thesis 1992:56, Trondheim.

Brassel, K.E. and Weibel, R., 1988, A review and conceptual framework of automated map generalization, *International Journal of Geographical Information Systems*, **2**(3), 229–44.

Bruegger, B.P. and Frank, A.U., 1989, Hierarchies over topological data structures, *Technical Papers 1989 ASPRS/ACSM Annual Convention*, Baltimore, 2–7 April, Vol. 4, pp. 137–45.

Buttenfield, B.P. and Delotto, J.S., 1989, Multiple representations, National Center for Geographic Information and Analysis, *NCGIA, Scientific Report for the Specialist Meeting, Technical Paper 89-3*.

De Marco, T., 1979, *Structured Analysis and System Specification*, New Jersey: Prentice-Hall.

Espoo City, 1988, TEKMA-Information Research, Final Report (in Finnish: TEKMA-tietojärjestelmätutkimus), Espoo, Finland.

Jones, C.B., 1991, Database architecture for multi-scale GIS, *Technical Papers 1991 ACSM-ASPRS Annual Convention*, Baltimore, Vol. 6, pp. 1–14.

Kilpeläinen, T., 1992, Multiple representations and knowledge-based generalization of topographical data, *Int. Arch. of Photogrammetry and Remote Sensing*, pp. 954–64, Washington, DC: Comm. III(B3).

Kilpeläinen, T. and Sarjakoski, T., 1993, Knowledge-based methods and multiple representation as means of on-line generalization, in *Proceedings of the 16th International Cartographic Conference*, Cologne, 3–9 May, Vol. 1, pp. 211–20.

Ledgard, H., 1983, *ADA, An Introduction*, 2nd Edn, New York: Springer-Verlag.

Mark, D.M., 1991, Object modelling and phenomenon-based generalization, in Buttenfield, B.P. and McMaster, R.B. (Eds) *Map Generalization: Making Rules for Knowledge Representation*, pp. 103–18, London: Longman.

McMaster, R.B. and Shea, K.S., 1992, *Generalization in Digital Cartography*, Washington, DC: Association of American Geographers.

National Center for Geographic Information and Analysis, NCGIA, 1989, The Research Plan of the National Center for Geographic Information and Analysis, *International Journal of Geographical Information Systems*, **3**(2), 117–36.

Robinson, A., Sale, R. and Morrison, J., 1978, *Elements of Cartography*, New York: John Wiley.

Shea, K.S. and McMaster, R.B., 1989, Cartographic generalization in a digital environment: when and how to generalize, *AUTO-CARTO 9, Proceedings of the 9th International Symposium on Computer-Assisted Cartography*, Baltimore, 2–7 April, pp. 56–67.

Ward, P.T. and Mellor, S., 1985 and 1986, *Structured Development for Real-Time Systems*, Vols 1–3, New York: Yourdon Press.

16

Experiment on formalizing the generalization process

Dan Lee

Mapping Sciences Division, Intergraph Corporation, 2501 Mercator Drive, Reston, VA 22091, USA

16.1 Computer-assisted approach

There have been discussions on the conceptual models of digital cartographic generalization by different researchers (Brassel and Weibel, 1988; McMaster and Shea, 1992). Presently, the implementation of a fully automated system for generalization is not practical. The fundamental task in developing such a system is to translate the conventional (manual) experience of map generalization to a computer process. This cannot be achieved without a complete understanding of how a cartographer handles the art and science of generalization.

An alternative way of pursuing digital map generalization is to start with an interactive system. Based on a decade of research and experiences on the development of a map generalization system for a special US government project, Intergraph Corporation has recently released a commercial product for cartographic generalization. This commercial product, known as MGE Map Generalizer (MGMG), provides a variety of generalization operators and visual tools for comprehensive generalization tasks. The operators automatically produce generalized output features, e.g., aggregated boundaries from a group of areal features or point features, simplified or smoothed linework, selectively reduced feature density and distribution, etc. (Lee, 1992). Human involvement during generalization includes choosing operators and algorithms, setting parameters and constraints, selecting features to generalize, and judging the results. This human/computer approach has been recognized by some researchers as 'amplified intelligence' (Weibel, 1991).

The digital map generalization system MGMG is intended to replace time-consuming and fully operator-dependent manual map generalization. The interactive approach gives the user the most flexibility and control on what to generalize, how to generalize, and how much to generalize. The experiment presented below is an example of formalizing a generalization process to achieve a target output.

16.2 Essence and concerns of generalization

Building knowledge for digital cartographic generalization involves the examination and interpretation of the essence of manual generalization processes. The challenge is to translate the human operations into a set of explicit, well-defined steps for computer implementation.

16.2.1 Interpretation of human process

Map generalization is known as a very subjective operation in conventional mapping, although it is constrained by the output scale and representation resolution. To achieve a successful generalization in the digital world, the entire process must be a simulation of the cartographer's visual analysis, decision-making, and actions. These three aspects form the main frame of the knowledge base for generalization.

What a cartographer learns and thinks while visualizing a map for generalization are (Lee, 1993):

- the characteristics of features (location, shape, size, orientation, spacing, etc.),
- the features densities (clusters vs sparseness),
- the distribution pattern of features (regular vs irregular), and
- the relationship among features (importance, priorities, and dependency).

Comparing the data to pre-defined criteria for output, the cartographer then decides where generalization should take place. Overcrowded areas and unimportant details are identified. The impacts and consequences of generalization actions (if a feature is generalized, how its related features are affected) are predicted.

Based on the above analysis and decisions, the generalized map is then created. Smoothed lines, combined boundaries, adjusted feature positions, and representative patterns are produced. As the cartographer is drawing the new map, he/she can easily achieve the objective of generalization, that is to reduce data complexity for the target scale and maintain the characteristics of geographic reality represented by the map, without violating the cartographic specifications.

The review of manual map generalization and the effort of interpreting the human mind led to the establishment of decomposed, well-defined steps for digital processes, and, therefore, the development of corresponding tools, functions and workflows to accomplish these steps.

16.2.2 Formalization of generalization process

With an interactive system, such as MGMG, the formalization of the generalization process is a learning process carried out between the user and the software. Part of the data analysis and decision making needs to be performed visually, for instance to find a heavily-curved linear feature to generalize or to judge whether a result is acceptable. Another part is buried in the algorithms, e.g. the cluster analysis and the determination of critical points.

Manual generalization actions can be considered to consist of the following nine categories of operations which become the major functions in MGMG design:

selection/elimination	aggregation
simplification	collapse
typification/refinement	exaggeration
classification/symbolization	displacement
aesthetic refinement	

Full definitions and descriptions of the above functions are given in a previous paper (Lee, 1993). Some of the functions have been implemented.

To formalize a generalization workflow using the existing product with the available functions and tools, the aspects discussed below should be taken into account.

Feature class dependency

Each feature class, for example hydrography or vegetation, has its special characteristics and, therefore, requires a unique generalization sequence.

Feature dimension dependency

Within a feature class, there are usually different dimensions of features (i.e. point, linear, and areal). Most of the generalization operators are designed for certain feature dimension(s). For example, simplification is for linear features (including areal boundaries) and aggregation is for points and areas. Therefore, the workflow for a feature class can be subdivided based on feature dimension.

Operational logic

It would be logical and efficient to start cartographic generalization with feature selection or elimination. Displacement and aesthetic refinement naturally follow most of the other operations. The intermediate steps vary with different cases. An iteration of operations is often necessary.

Minimizing spatial accuracy reduction

Digital generalization results in changes or displacement of data points from their original positions. Part of the displacement can be minimized by, besides carefully choosing algorithms and parameters, executing generalization operations in a certain order.

Maximizing process efficiency

In cases where processing efficiency is important (as in the case of large data volumes or where the scale change is drastic), the workflow should be customized in order to maximize throughput.

Minimizing the need for subsequent readjustment

Unnecessary repetitions of the same operation should be avoided. For example, to generalize a cluster of small areas, aggregation followed by smoothing may be better than smoothing followed by aggregation, because the latter may require further smoothing of the aggregated boundaries.

16.3 Experimental

True knowledge comes from practice. An experimental project was carried out to establish a potential workflow for generalization of topographic data. MGMG Version 1.0 (released in July 1993) was used. An incomplete set of USGS digital topographic data, which contains vegetation, hydrography, individual buildings, and developed areas (see Figure 16.1), was used. Unfortunately, the vegetation and developed areas are in the same shade of gray due to the limitation of the printing device. The source map is the Herndon, VA quad sheet at 1:24 000 scale (Figure 16.1 is a reduced plot from the original scale for illustration purposes). The target scale is 1:100 000. Due to the lack of documented specifications for the output, the existing hard copy of the final product (the map) was used as the reference for determining the amount of information to be represented at the reduced scale.

The workflow reflected the considerations discussed earlier. A series of intermediate results are presented in Figure 16.2 to Figure 16.9. Most of the intermediate results for vegetation data are shown, while only a couple of significant intermediate results for other features are given due to the limited chapter length. Figure 16.10 is the final presentation of the experimental result. The corresponding operational sequences and criteria are explained below.

Workflow for vegetation data generalization (see Figure 16.2 for the original presentation):

- **Elimination**
  Small and unimportant areas were eliminated (Figure 16.3).
- **Area-to-line collapse**
  Areas that are too narrow and insignificant were found and reduced to linear features (Figure 16.4).
- **Elimination**
  The linear features created from the above step were eliminated by length or feature type. The small areas generated from the last step were eliminated (Figure 16.5).
- **Area aggregation**
  Areas in clusters were combined into larger areas (Figure 16.6).
- **Simplification**
  New areal boundaries were simplified locally to represent the features with the minimum amount of data.
- **Smoothing**
  The simplified boundaries were smoothed locally for aesthetic quality.
- **Comparison and adjustment**
  The result was compared with the real map at 1:100 000 scale. Minor editing/adjustment was involved to compensate the process during generalization and to refine the final output (Figure 16.10).

Workflow for hydrography data generalization (see Figure 16.1 for the original presentation):

- **Elimination**
  Very short branches, very small lakes, and unimportant features were eliminated.
- **Area-to-line collapse**
  The lake in the middle of a river was partially collapsed to become part of the river (Figure 16.7).

Figure 16.1 Original presentation of USGS DLG data of Herndon, VA at 1:24 000 containing hydrography, vegetation, individual buildings, and developed areas. It is photo-reduced for illustration.

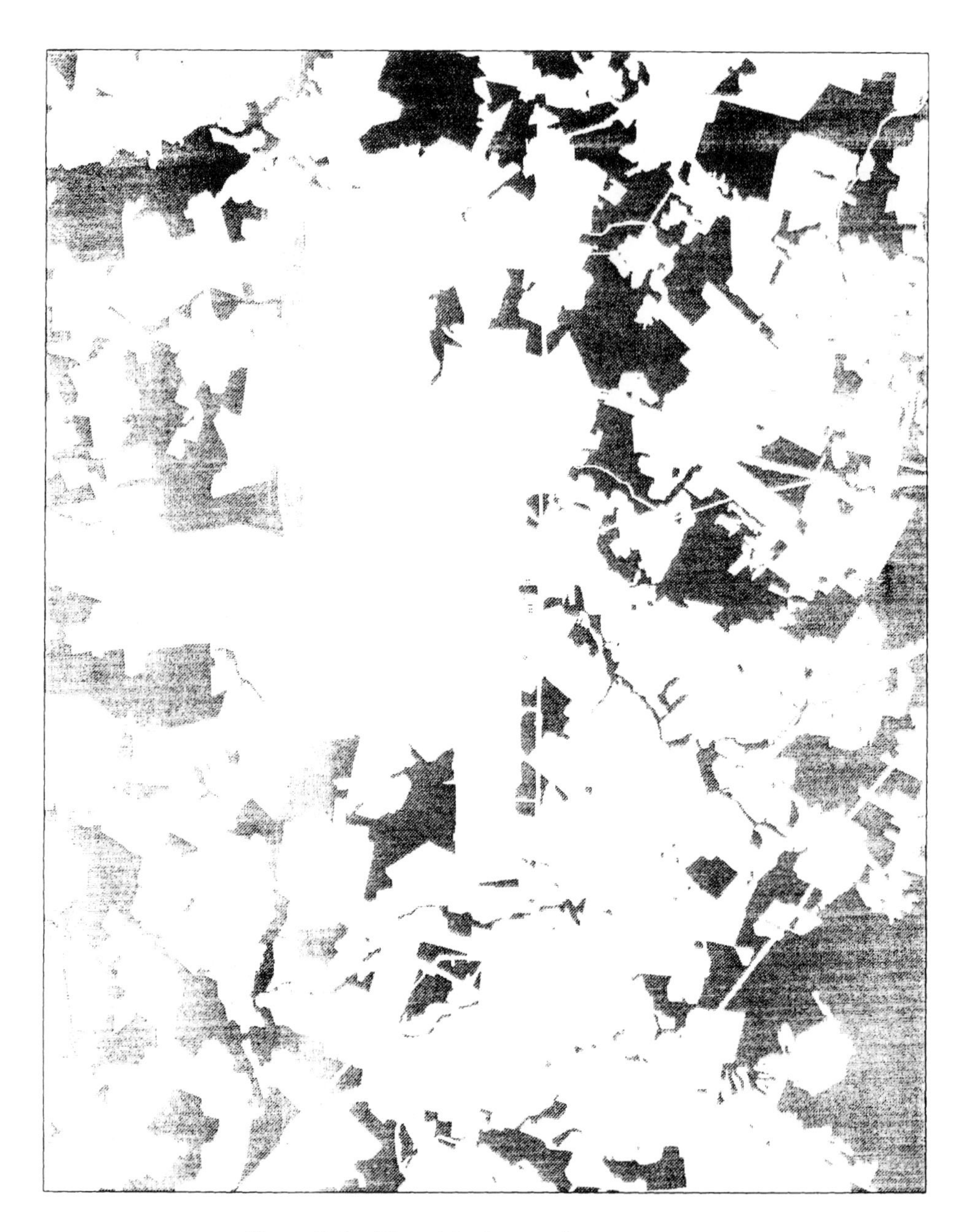

Figure 16.2 The original vegetation presentation.

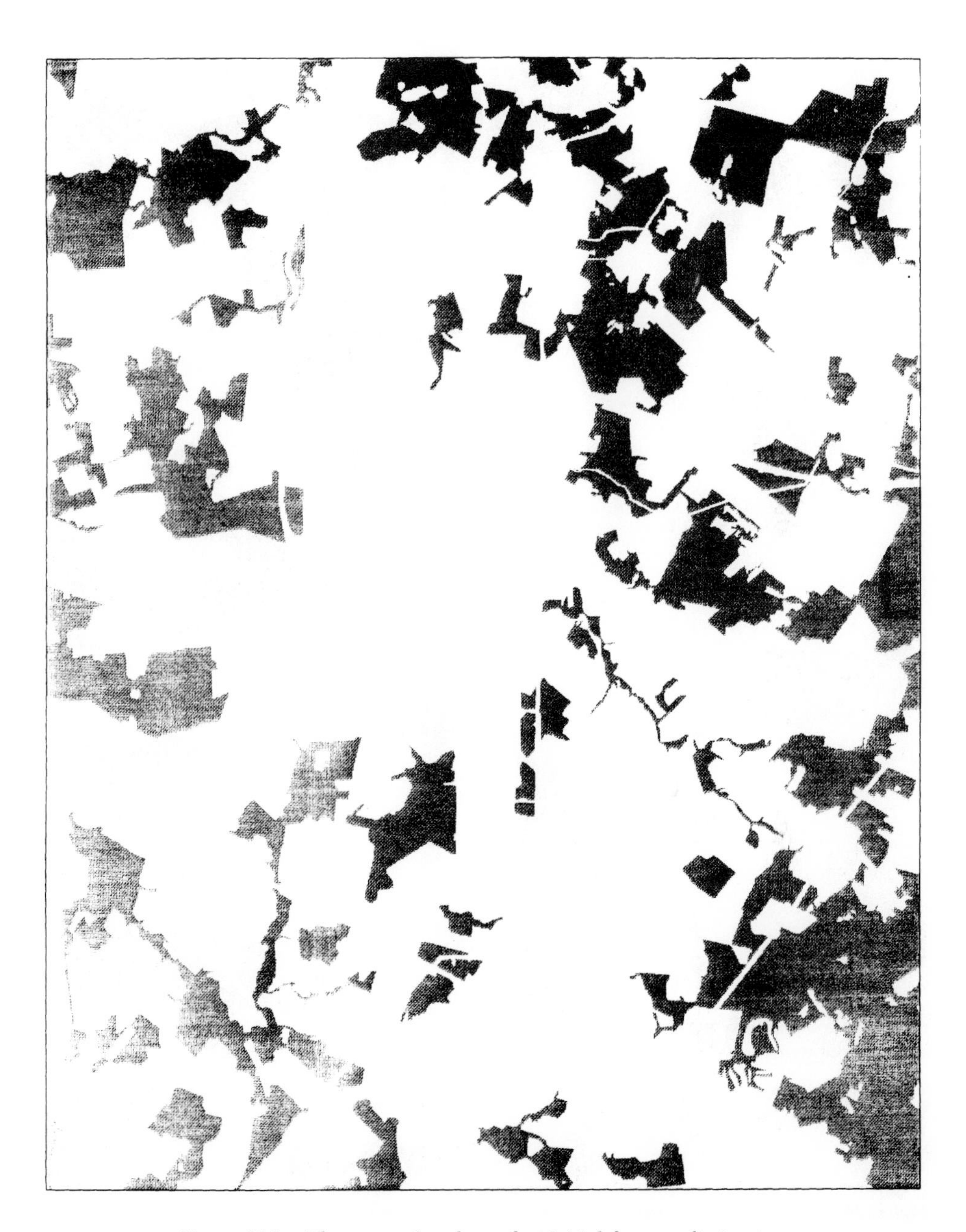

Figure 16.3 The vegetation data after initial feature elimination.

Figure 16.4 Areas that are too narrow were found using area-to-line collapse and were reduced to linear features.

Figure 16.5 Further elimination removed the linear features shown in Figure 16.4 and other unimportant features.

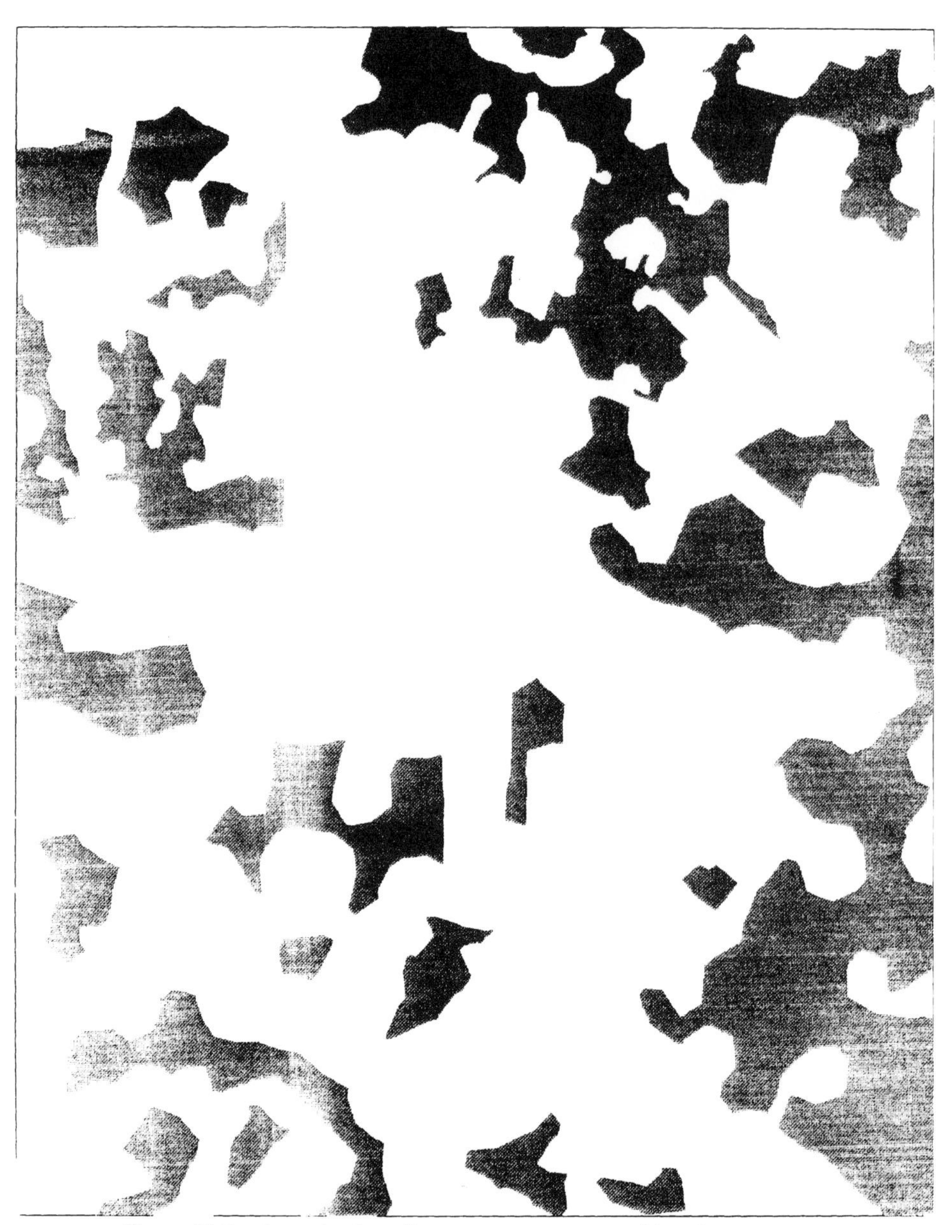

Figure 16.6 Areas in close distance were aggregated into larger areas.

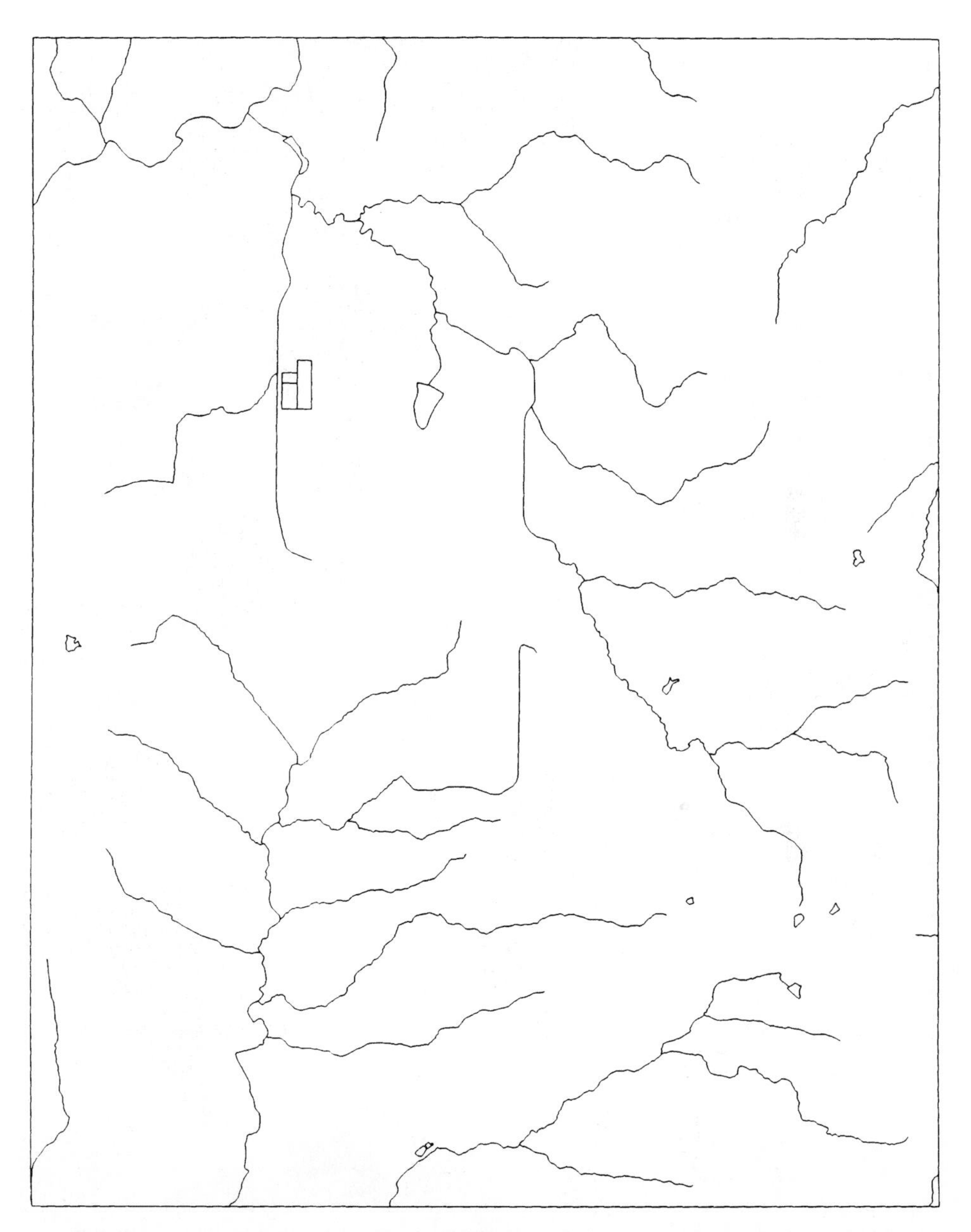

Figure 16.7 Hydrographic data after initial feature elimination and area-to-line partial collapse.

Figure 16.8 Individual buildings and developed areas after initial feature elimination.

Figure 16.9 Aggregated (points and areas) and simplified data from individual buildings and developed areas.

Figure 16.10 Final result for the scale of 1:100 000.

- **Simplification**
  The complexity and the amount of data representing the hydrography was reduced.
- **Smoothing**
  The simplified linear features were smoothed for aesthetic quality.
- **Comparison and adjustment**
  The result was compared with the real map at 1:100 000 scale. Minor editing/adjustment was needed to compensate the process during generalization and to refine the final output (Figure 16.10).

Workflow for individual buildings and developed areas (see Figure 16.1 for the original presentation). Since the same output features, urban areas, were produced from both of these features, the workflow consists of a separate process for each feature type and a combined process for final output.

For individual buildings:

- **Elimination**
  Unimportant building points were eliminated (Figure 16.8).
- **Point aggregation**
  Point clusters were changed into areas.
- **Area aggregation**
  Area clusters generated from the last step were further aggregated into larger areas.
- **Elimination**
  Small or unwanted areas from the above steps were removed.
- **Simplification**
  The new area boundaries were simplified to reduce the amount of data for further processing (see the black areas in Figure 16.9).

For developed areas:

- **Elimination**
  Small and unimportant areas were eliminated (Figure 16.8).
- **Area aggregation**
  Area clusters were combined into larger areas.
- **Elimination**
  Unwanted areas from the above steps were removed.
- **Simplification**
  The new area boundaries were simplified to reduce the amount of data for further processing (see the gray areas in Figure 16.9).

Combined process:

- **Area aggregation**
  The areas generated from two different source feature types were combined into the same output area features.
- **Simplification**
  The new areal boundaries were reduced to fewer points.
- **Comparison and adjustment**
  The result was compared with the real map at 1:100 000 scale. Minor editing/adjustment was applied to compensate the process during generalization and to refine the final output (Figure 16.10).

Finally, the three new feature types were displayed simultaneously. Some minor editing was carried out to ensure feature relationships and to improve the visual impression of the output. The result, Figure 16.10, was printed at approximately 1:100 000 scale. The amount of information kept in the final output was fairly comparable to the real map.

From this initial experiment, it was found that setting parameters for certain feature types and feature groups were the most critical step. The powerful dynamic display capability made this task easy. Different operational sequences resulted in different output and the 'Undo' command made the trial–error learning process possible.

16.4 Conclusions

Although the experiment was preliminary, the outcome was quite encouraging. This experiment proves that an interactive system, such as MGMG, is very helpful for knowledge acquisition for spatial data generalization. It gives the user control over the degree of generalization and the flexibility of comparing different effects when changing variables. The parameters were saved and further adjustments could be carried out to refine the process, improve the result, and optimize the workflow.

The interactive process may still reflect the subjective nature of generalization. However, the processing time and consistency were considered significantly superior to manual generalization. Digital map generalization is certainly finding more and more applications in GIS areas, in addition to cartography.

References

Brassel, K.E. and Weibel, R., 1988, A review and conceptual framework of automated map generalization', *International journal of geographical information systems*, **2**(3), 229–44.

Lee, D., 1992, 'Cartographic Generalization', Intergraph Internal Technical Paper.

Lee, D., 1993, From master database to multiple cartographic representations, in *Proceedings of the 16th International Cartographic Conference*, Cologne, Germany, 3–9 May, Vol. 2, pp. 1075–85.

McMaster, R.B. and Shea, K.S., 1992, *Generalization in Digital Cartography*, Washington, DC: Association of American Geographers.

Weibel, R., 1991, Amplified intelligence and rule-based systems, *Map Generalization: Making Rules for Knowledge Representation*, pp.172–86, London: Longman.

17

A hierarchical top-down bottom-up approach to topographic map generalization

Gary J. Robinson

NERC Unit for Thematic Information Systems, Department of Geography, University of Reading, Reading RG6 2AB, UK

17.1 Introduction

Generalization is one of the most important factors in the efficient and effective visualization of spatially referenced data. This is particularly true of data held in a Geographic Information System (GIS), although some would argue that the ability of GIS to zoom in 'indefinitely' means that generalization in the cartographic sense is not required. Such views reflect a widespread lack of appreciation of the limitation of computers. After all, computers can only do what they are told: whether the results are meaningful is another matter. The use of inappropriate scales, and especially the combination of data from different source scales, frequently leads to misinterpretations, or at least unwarranted expectations, of the results of GIS operations. The onus is therefore on the user to ensure that the data are at the appropriate scale for the problem in hand, which has implications on accuracy, precision and quality (Goodchild, 1991).

The relationship between scale and accuracy is especially important at the data capture stage, where all sorts of generalization may (and usually) have been carried out on the source material. Clearly, it is pointless digitizing to a given accuracy level if the scale of the data is too small. Some subtle errors in GIS operations arise because of the strong context-sensitive nature of generalization. Where the source datasets for a particular application have been generalized in tandem there is not a problem: spatial relationships between features can easily be maintained in changing scale. In contrast, with data from different sources that have been generalized independently (despite being at the same original scale), there are no such guarantees. This is one area where, for example, overlay operations can create spurious effects that have to be corrected, generally using post-processing operations. Since the comparison and combination of diverse data sources is the 'raison d'être' of GIS, the importance of generalization cannot be underestimated, and certainly not ignored.

The recent upsurge of interest in generalization by commercial GIS vendors such as Intergraph is therefore welcome, although this does raise some questions. Is this interest user driven, or are the vendors waking up to the deficiencies of their products and/or seeing new markets? Where can the academic community best target its efforts? And how should the three groups work together? Many other questions also have to be answered before research into automated generalization can progress. For example, which generalization operations really do work satisfactorily on a computer and which do not? What, if any, changes need to be made to spatial data structures? Just what are the roles for expert systems and object-oriented approaches? Is batch mode generalization feasible, or will some degree of user interaction always be necessary?

This chapter attempts to address some of these issues and others raised in the position paper on generalization (Chapter 1), using the development of an automated topographic generalization system for digital large-scale maps from the Ordnance Survey of Great Britain (OSGB) as a case study.

17.2 Generalization

17.2.1 Definition of generalization

In the mapping context, generalization can be defined as the process of reducing the amount of detail so that the character or essence of the original features is retained at successively smaller scales. This is certainly true of topographic map generalization, which is a purely cartographic exercise involving omission, aggregation, simplification, displacement, exaggeration and symbolization of either individual features or groups of features.

Generalization in the geographical sense is much broader, since it encompasses the representation and treatment of physical, social, economic and other spatially referenced processes at different scales. The use of conceptual 'models' to describe such processes is becoming widespread, and it is natural to investigate whether these ideas can be applied in the field of cartographic generalization.

17.2.2 Manual generalization

An extensive body of literature, consisting of many rules and guidelines, which describes the generalization operations employed in manual map production has been drawn up over many decades. Within Great Britain for example, the Ordnance Survey has gathered the guidelines it uses in-house into a series of formal drawing instructions for surveyors, draughtsmen and cartographers. Similar publications no doubt exist in many other mapping agencies throughout the world. Other guidelines on manual generalization techniques (of which topographic mapping is one component) have appeared in scores of books over the years, some specialized in nature, others more general.

Even a casual analysis of the literature reveals two things. First, the 'art' of manual generalization is a well-established albeit complex topic. Second, despite the apparent depth and diversity of these 'guidelines', the nature of this information is, to say the least, 'fuzzy' and, possibly more worryingly, incomplete. These observations indicate that in order to develop a successful automated generalization system, considerable effort must be expended in firming up these rules and filling in the gaps — possibly through a series of 'knowledge-elicitation' exercises, which are widely used in the development of expert systems. This is likely to be an interesting exercise, since the experts concerned often admit that they find it difficult to rationalize their decisions into formal rules (Chapter 1).

17.2.3 Early attempts at automated generalization

Until recently, investigations into how such knowledge can be transferred to the digital domain have tended to concentrate on relatively small subsets of the generalization process. Simplification of linear features dominates the early literature (Lang, 1969; Douglas and Peucker, 1973; Jenks, 1981). In contrast, relatively few methods for displacing cartographic features and generalizing areal features have been developed (Lichtner, 1979; Monmonier, 1983; Meyer, 1986).

Most research into automated generalization has also tended to be at small scales. There are several reasons for this. Maps become relatively simpler, in terms of the variety of features they contain, with decreasing scale. Another reason arises from the different dominant types of features at each scale. Large-scale maps, especially in urban areas, consist mainly of areal objects such as buildings, land parcels and road surfaces, whereas smali-scale maps contain mainly linear and point features such as contours, rivers, (stylized) roads and point symbols. The form of features and their spatial relationships with one another also tend to be simpler at smaller scales. This often allows features to be treated independently during generalization. In contrast, the considerably more complex spatial interactions at large scales require the adoption of more sophisticated approaches.

17.2.4 Current attempts at automated generalization

Whilst this early work concentrated on the generalization of individual features, there has been a noticeable shift in recent years towards investigating the relationships amongst multiple features, or 'objects'. As a result, some interesting 'knowledge-based' models have been proposed (Nickerson and Freeman, 1986; McMaster, 1991; Weibel, 1991). These have tended to be rule-based systems, in which the rules have been derived from existing manual operations, or empirically through experimentation.

The generalization process has only really been tackled in a piecemeal manner, with varying degrees of success. Those developments that have emerged have tended to be rather limited and invariably application dependent. However, this probably reflects the complexity of the spatial relationships amongst map features — plus a corresponding lack of suitable data structures — rather than a lack of coherent strategies based on models of the underlying processes. A particular difficulty is in attempting to model the parallel nature of the generalization process. Despite the apparently simple concepts of the basic tasks involved (simplification, displacement, etc.), these are generally applied in a parallel fashion, or at least in a highly iterative manner, in the manual domain. Computers are not very good at doing this: they do not see the overall picture and the interrelationships within it. Instead, they tend to concentrate on small, well-defined, low-level tasks. The limited successes in automated generalization have therefore tended to be in those areas where parts of the problem can be considered 'separable', e.g. line simplification at medium and small scales.

The fundamental basis of generalization at a theoretical level has recently been questioned (Brassel and Weibel, 1988). This highlights the growing belief, at least in the academic community, that generalization is still not understood well enough to enable completely automated solutions to be developed, and that the entire process should be looked at carefully — in effect 'taking a few steps back' from the problem.

17.3 Is there a need for automated generalization?

Unquestionably, yes. There is an urgent need for practical solutions now, given the ever increasing growth of fully digital map bases in many countries. Much of these data are generated in a form suitable for GIS/LIS-based applications, which require accurate and timely information.This is not to suggest that this is the only market: the main output of many national mapping agencies is still the standard map series, and this has implications on how, when and where automated generalization techniques should be introduced into existing production mapping flowlines. Certainly, however, the rapid growth of GIS is the major driving force in calls for suitable generalization tools to be developed.

The other aspect that should be investigated as a matter of urgency is the content, form and quality of the actual output from GIS, both on screen and on paper. Leaving cartographic design in the hands of the (naive) user is asking for trouble!

17.3.1 Scale-free or fixed-scale series?

An important question about where automation should be introduced is whether scale-free generalization is required. This is clearly the case for applications such as GIS or LIS, which produce ephemeral or 'on-the-fly' maps on screens and slightly more permanent paper output. In contrast, most paper map series are produced at a few fixed scales. It is this latter case that has usually influenced past attempts at introducing 'complete' automated generalization systems, e.g. CHANGE (Powitz and Schmidt, 1992). The reason is fairly clear: there are existing rules to work from. Whether these are sufficient is the subject of the discussion on the case study. By removing the constraint of fixed scales, such rules would have to be generalized. As these are often specific, or to use expert-system terminology 'non-fuzzy', this may be more difficult than starting from scratch.

17.3.2 Paper versus VDU

Another factor that should be borne in mind is the nature of the output medium, since this has an impact on the amount and type of information that can be optimally presented to the user. Paper maps are generally acknowledged to be capable of holding more information per unit area than VDU-based ones. VDU maps, however, are more flexible: selected categories of data can be omitted or highlighted very easily, and they can be updated more quickly.

Whether paper maps will still be used as input to GIS in the future is another, and hopefully temporary, issue. The information on them is compiled and generalized to suit the paper medium and the particular requirements of that medium. As mentioned earlier, digitizing such products for GIS applications is fraught with danger, and is often counter-productive since the original information may already be in (ungeneralized) digital form.

17.3.3 Update

Changes in the landscape, usually the built environment, invariably lead to changes in the basic scale digital data. How these changes are transmitted to derived smaller-scale datasets depends on whether a 'scale-free' or 'fixed-scale' approach is adopted. To some extent, this is less of a problem for scale-free data since the basic scale data are the ultimate reference. For derived fixed-scale maps, however, such changes can be more difficult to handle. What is required is consistency in the output from the generalization process, regardless of the order in which individual features are handled. This topic is discussed in the following case study.

17.4 Case study: OSGB large-scale maps

OSGB is currently establishing a comprehensive basic scale digital coverage of Great Britain. These scales are 1:1250 for urban areas, 1:2500 in rural areas and 1:10 000 in mountainous regions. Selected subsets (primarily elevation and communication layers) of derived and smaller non-derived scales (1:25 000, 1:50 000, 1:250 000 and 1:625 000) are also being digitized. The relationships between the current manually produced versions of these series are complex, with revision seemingly carried out independently at several scales, especially at smaller scales (Figure 17.1). The reason is somewhat historical: e.g. contours in the smaller-scale series were derived from Imperial (feet) contours from the old 'County' map series and have little or no link with their larger-scale metric equivalents.

Whether these scales should be fully linked, with a continuous well-controlled update flowline, is currently the subject of debate within OSGB. Were appropriate automated techniques already available, the matter might be settled relatively quickly. However,

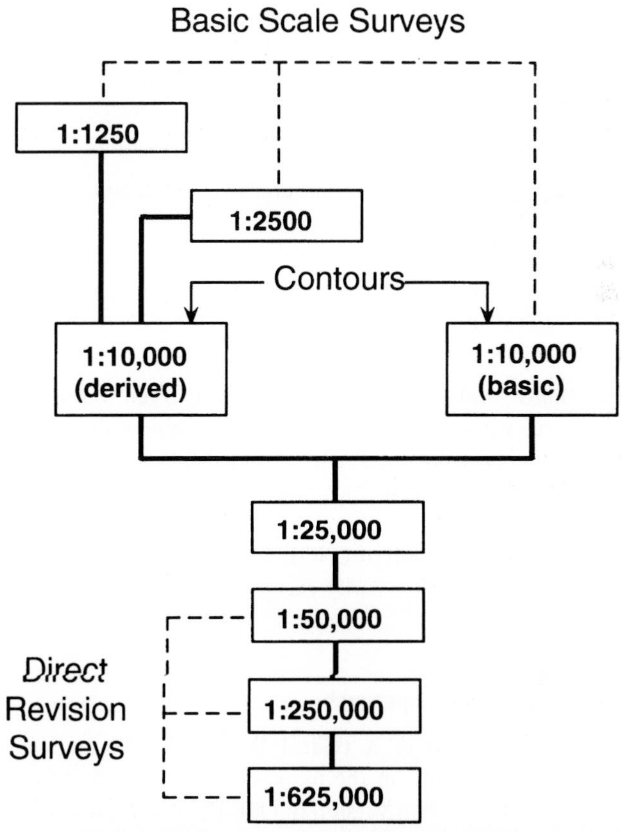

Figure 17.1 Revision scheme of Ordnance Survey map series. Thick lines denote derivation via (manual) generalization, where appropriate.

since they are not, the situation is somewhat dependent on such techniques being developed. OSGB has looked at automated generalization in the past and is presently reviewing the situation. They are probably not alone: mapping organizations in other countries are presumably in the same dilemma, waiting to see what emerges from the commercial and academic world. This impasse is having an unfortunate impact on the development of techniques, since until there is a sufficiently large demonstrable demand for automated products, national mapping agencies will not proceed along this route (and help to fund academics), but until such data are available, the end customers will not request them.

17.4.1 OSGEN 1: analysing OSGB generalization 'rules'

In a previous study (Robinson and Zaltash, 1989) the generalization rules used by OSGB in generalizing from large (1:1250 and 1:2500) to medium scale (1:10 000) were analysed and put into a rule-based expert system adviser called OSGEN (OS GENeralization). The reason for doing this was not to develop an automated generalization system *per se*, but rather to investigate the nature of the rules, the interactions among them, whether they could be transformed into a computer-understandable form, and, most importantly, how complete they were. The result of this study revealed that most of the rules were too specific and could not be applied as stated in many cases. The most disturbing finding was that the rules were not complete by any stretch of the imagination. Evidently, a considerable amount of cartographic knowledge is held in cartographers' heads. This reinforced the fact that these 'rules' are no more than guidelines, and are really intended as such, although there are many cases where they can be applied rigorously.

In terms of computer understandability, most of these rules were found to fall into two diametrically opposed categories: under-specific and over-specific. An example of the first category is a rule dealing with the treatment of a small building:

> 'if the building is remote and topographically significant, then it should be retained; otherwise, it should be omitted'.

In contrast, a rule such as:

> 'if the shortest side of the building is less than the specified tolerance, then it should be enlarged to the minimum size'

is very specific or 'hard', and considerable latitude was often found in its interpretation. This was especially true where the building lay close to other buildings and roads and was subject to other rules concerning aggregation and displacement. This interaction — or more accurately conflict — between rules, highlights the difficulty of accounting for all possible cases of spatial context in the guidelines.This led to the conclusion, after further experiments with a modified version of OSGEN (OSGEN 2), that a rule-based approach using these 'rules' would fail quite badly except under certain well-defined but limited situations.

17.4.2 OSGEN 3: a pragmatic approach

Fortunately, the generalization of a typical large-scale OS map can be considered 'limited', since most of the space on the map is occupied by buildings and roads. The approach being taken in the current version of OSGEN is extremely pragmatic, and relies on the relative importance of certain features over others. For example roads — or more strictly road edges — have priority over all other features, followed by buildings, and then fences. Since the first two categories comprise over 80 per cent of the total content of a typical urban map, it was thought that a significant part of the problem could be solved by dealing with these two categories. Only a brief overview of the approach is given here: more details appear elsewhere (Robinson and Lee, 1994).

The style of the current 1:10 000 scale OS map series is partly dictated by the production requirements of the smaller-scale 1:25 000 series (Figure 17.1). The degree of generalization is therefore slightly greater than would normally be applied, particularly in the case of buildings and roads. The availability of automated generalization tools would reduce, if not remove, this constraint, as well as reducing the update cycle of both series and making the 1:25 000 scale maps available in digital form.

Data and data structure

The basis of the data structure used in OSGEN 3 is taken from the OS topological data model (Brading, 1989) with a few modifications. This is a full topological vector-based model, which was designed for mapping and GIS purposes. It allows the basic graphical data and attribute information (feature codes, etc.) to be stored within and accessed from a relational database, in our case Oracle. The test area used in this study was part of central Birmingham, the data for which were supplied by OSGB as part of their contribution to this work.

'Divide and conquer'

The starting point for OSGEN 3 was to exploit the natural networks formed by roads and other communication features within the map. This enables features, as graphic elements, to be partitioned into convenient 'blocks' (right-hand side of Figure 17.2). The main reason for doing this is that, to a good approximation, only those features within a block interact with each other: interactions between features, even in adjacent blocks (e.g. lying on different sides of a street), are rare. This considerably simplifies the problem, reducing the amount of computing required for spatial comparisons and data retrieval. Another advantage is that update can be carried out on a block-by-block basis, with minimal impact on surrounding areas.

Objects, meta-objects and spatial relationships

After partitioning into blocks, physically meaningful 'objects' such as individual buildings and road edges are created from the feature-coded graphical features using a Prolog rule base. However, because the coding scheme adopted by OSGB does not differentiate sufficiently among object types, especially buildings, an additional level of processing is necessary. This infers the type of an object from its geometrical characteristics and neighbourhood relationship with nearby features. For example, houses are generally within a certain size range, have only a few corners, occur in groups, and are located next to roads.

Logically related objects such as terraced houses, factory complexes and roads are then aggregated into higher-level 'meta-objects', with spatial relationships within each meta-object held as attributes. This process is continued at successively higher levels until a complete hierarchy of objects and their relationships is generated (Figure 17.2). Blocks can also be treated as meta-objects since these can be assigned information about the general character of the local area, which is useful in generalization even at smaller scales,

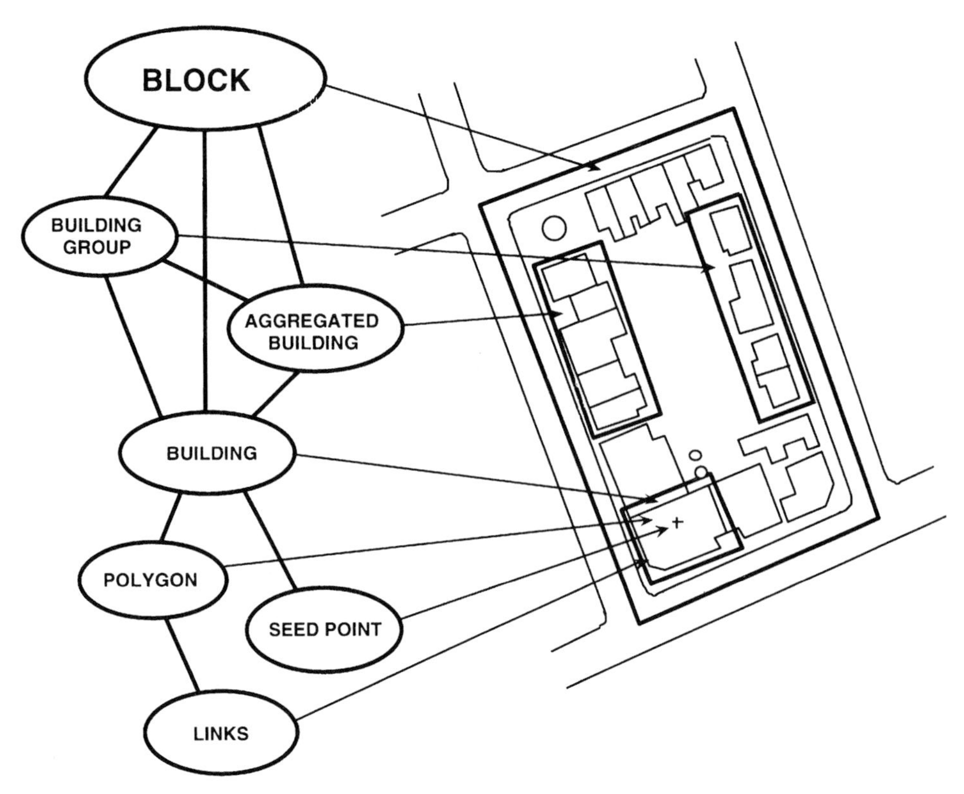

Figure 17.2 Bottom-up creation of (building) object hierarchy for top-down generalization. Information used in generalization activities, e.g. building type, area, orientation and neighbourhood relationships, is held at the appropriate level. It is generated from the attributes (feature codes) of the basic features and their geometrical properties, and propagated upwards as appropriate.

e.g. 1:50 000. The end result is a model of the map; whether this is a 'cartographic' model or an 'abstract physical' model is arguable.

Generalizing through deferred 'actions'

The advantage of this model-based approach is that the OSGB 'generalization rules' relate to these objects more readily than the original graphic-based features on the map. In particular, it is possible to carry out context-sensitive generalization of, for example, a group of houses in an estate that should end up looking more similar than if they were isolated. However, the OSGB 'rules' are not applied immediately, since the order in which they are applied can lead to different results depending on which object is generalized first (Armstrong, 1991). Instead, the rules are applied in a 'deferred' mode at the appropriate level to the meta-objects, objects, and features in the hierarchical model (Figure 17.3). This results in a set of zero or more potential 'actions' for each object. Each set is then ordered using a list of priorities held in a rule base. If the action with the highest priority is to omit the object then this is done. Otherwise, where there is more than one action a conflict resolution process is invoked, again using a rule base. Some actions can be combined: e.g. a building may be required to be shifted east and north simulta-

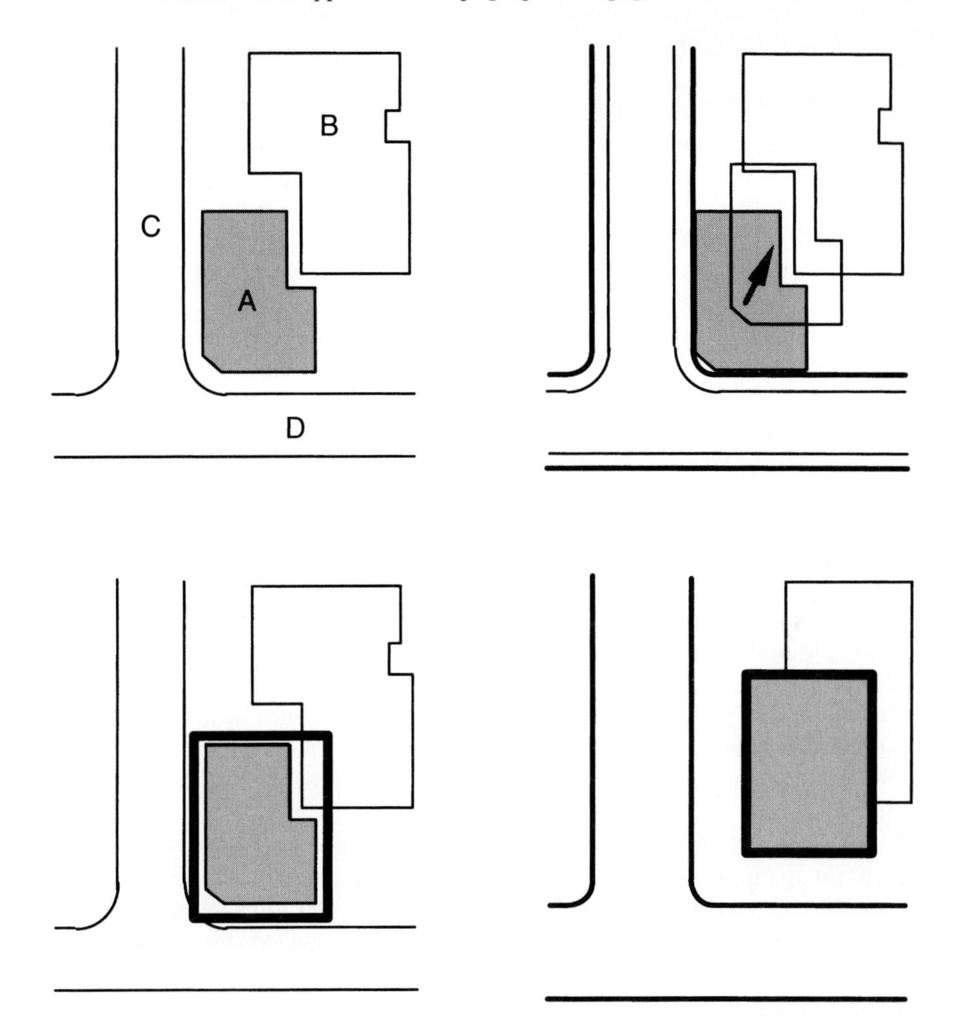

Figure 17.3 Resolution of conflicts when applying generalization rules to a (public) building. Top left: original data showing public building (A) in relation to surrounding roads (C and D) and a nearby building (B). Top right: enlargement of C and D would displace A in north and east directions to maintain adequate space between them, leading to clash with B. Bottom left: simplification and exaggeration of A causes additional clashes with original C, D and B. Bottom right: result of using 'deferred actions' to generalize A, B, C and D 'simultaneously'.

neously. If there are no further conflicting actions, such as exaggerating the building, then these displacements are combined. This process carries on iteratively until one or more non-competing actions result. If conflicting actions result, e.g. all the sides of a building have to be shifted inwards — violating another action that is to portray the object at the minimum size — then higher-level conflict-resolution rules come into play. If necessary, the problem can be flagged and the decision left to a human operator.

The advantage of this method is that the generalization rule base is considerably simpler, avoiding the 'combinatorial explosion' that frequently results with attempting to cater for every possible case that could occur. The idea of 'action conflicts', in particular, allows greater differentiation between those problems that can be handled automatically and those that cannot.

17.5 Conclusions

Preliminary results from the OSGEN 3 study show that a model-based approach is appropriate for large-scale topographic map generalization. In particular, the technique of 'deferring' generalization actions appears relatively successful in solving the conflicts that occur when they are applied in a different order, and to some extent the parallel nature of their operation on multiple features.

This study also indicates that for this type of map over 80 per cent of the generalization process can be carried out automatically, and that it is possible to identify correctly those that require user intervention. This is important for mapping agencies that require as near to a fully batch-oriented generalization procedure as possible, but who are willing to tidy up a small proportion by hand.

Existing written rules intended as guidelines in manual generalization are generally unsuitable for incorporation into expert-system-based automated generalization systems as they stand. The vaguer of these rules need qualifying clauses added to enable them to be manipulated by computers, while the more specific ones need to be expressed in more general terms, i.e. they need to be 'generalized'. Looked at another way, such rules are perhaps not a good starting point: it may be better to look at the underlying theory behind why and how they were created in the first place. This would also have the benefit of enabling specialized products at non-standard scales to be created.

Whether the approaches used in the OSGEN 3 study can be applied to more general situations requires further investigation. However, the basic idea of using a generalization-specific hierarchical model of spatially referenced information appears promising. A model-based approach also suggests that object-oriented programming techniques merit investigation.

The case study described here is a good example of how a mapping agency and an academic group can work together to produce a 'proof of concept' demonstration system with some useful working elements. That the cooperation of a software vendor would be required to turn this into a more efficient working system highlights the need for all three parties to collaborate, especially in the early stages.

As a general point, it is acknowledged that a detailed description of the generalization process, coupled with appropriate models of the spatial phenomena being studied, are essential for automated generalization techniques to work properly. All too often one sees the same simplification algorithm applied to totally different types of linear feature, with similar outcomes; a context-sensitive approach is clearly called for. However, perhaps the most critical issue is the underlying data structure, both in terms of the representation of spatial entities and their interrelationships, and the holding of contextual and generalization-specific higher level information. Without this, the full integration of individual tools such as simplification, aggregation, and displacement will remain elusive.

Acknowledgements

This chapter was produced as part of a general research topic under NERC contract F60/G6/12. The automated topographic generalization research project discussed in this chapter was funded jointly by the Great Britain Ordnance Survey and Reading University.

References

Armstrong, M.P., 1991, Knowledge classification and organization, in Buttenfield, B.P. and McMaster, R.B. (Eds) *Map Generalization: Making Rules for Knowledge Representation*, pp. 86–102, London: Longman.

Brading, G., 1989, Topographic database relational data structure, Version 2.3, Ordnance Survey Internal Document, October, 25 pp.

Brassel, K.E. and Weibel, R., 1989, A review and conceptual framework of automated map generalization, *International Journal of Geographic Information Systems*, **2**(3), 224–9.

Douglas, D.H. and Peucker, T.K., 1973, Algorithms for the reduction of the number of points required to represent a digitised line or its caricature, *Canadian Cartographer*, **10**(2), 112–22.

Goodchild, M.F., 1991, Issues of quality and uncertainty, in Müller, J.C. (Ed.) *Advances in Cartography*, pp. 113–39, Barking, Essex: Elsevier.

Jenks, G.F., 1981, Lines, computers and human frailties, *Annals of Association of American Geographers*, **71**(1), 1–10.

Lang, T., 1969, Rules for robot draughtsmen, *Geographical Magazine*, **62**(1), 50–1.

Lichtner, W., 1979, Computer-assisted processes of cartographic generalization in topographic maps, *Geo-Processing*, **1**, 183–99.

McMaster, R.B., 1991, Knowledge acquisition for cartogrpahic generalization, in *Proceedings of the 15th Conference of the International Cartographic Association*, Bournemouth, UK, 24 pp.

Meyer, U., 1986, Software-developments for computer-assisted generalization, in *Proceedings of Auto-Carto London*, 1986, **2**.

Monmonier, M., 1983, Raster-made area generalization for land use and land cover maps, *Cartographica*, **20**(4), 65–91.

Nickerson, B. and Freeman, H., 1986, Development of a rule-based system for automated map generalization, in *Proceedings of the 2nd International Symposium on Special Data Handling*, Washington, DC, 5–10 July, pp. 537–56.

Powitz, B.M. and Schmidt, C., 1992, 'CHANGE', Internal Technical Report, Institute for Cartography, University of Hannover, 16 pp.

Robinson, G.J. and Lee, F., 1994, An automated generalization system for large scale topographic maps, in Worboys, M.F. (Ed.) *Innovations in GIS*, Vol. 1, Selected Proceedings from the First National Conference on GIS Research UK, pp. 53–63, London: Taylor & Francis.

Robinson, G.J. and Zaltash, A., 1989, Application of expert systems to topographic map generalisation, in *Proceedings of the Conference of the Association of Geographic Information*, Birmingham, pp. A3.1–3.6.

Weibel, R., 1991, Specification for a platform to support research in map generalisation, in *Proceedings of the 15th Conference of the International Cartographic Association*, Bournemouth, UK, 12 pp.

18

Strategies for ATKIS-related cartographic products

Georg Vickus

Landesvermessungsamt Nordrhein-Westfalen, Muffendorfer Strasse 19–21, 53177 Bonn, North-Rhine-Westphalia, Germany

18.1 Introduction

Cartography is considered as the science of describing and documenting the earth's surface by collecting knowledge on its structure and relief, called geoinformation, providing it to public and private users as standardized digital spatial data and analogue maps, and looking for the legal aspects of property on land as well. It includes the definition and measurement of a geodetic reference system, the specification of earth- and map-related coordinate systems, photogrammetric, topographic and remote sensing techniques for collecting data about the objects and phenomena on the earth's surface, their documentation in Digital Image Models, Digital Landscape Models and Digital Cartographic Models, or in an analogue manner, and the derivation of digital and analogue products (maps and charts) in a wide variety of content, scales and graphical representations. The geoinformation has to be prepared for application fields like Survey; Geology; Environmental Protection; Ground, Water and Air Navigation; Agriculture and Forestry; Hydrology and Hydrography; Cadastre; and Geophysics; in order to meet the wide variety of users' requirements.

Within this common context of cartography, it is the legal task of the official Surveying and Mapping Agencies to collect, document and provide basic land-related data on the earth's surface based on clear defined standards. In addition, they have to produce, print, edit and distribute the official topographic maps and look after the state's interests according to the use of maps and digital data by third parties.

To fulfil the expected needs on digital cartography, the Surveying and Mapping Agencies in Germany developed the Digital Cadastral Map ALK (Automatisierte Liegenschaftskarte) and the Authoritative Topographic–Cartographic Information System (ATKIS) as a standard, and decided on the building of ATKIS databases until 1995 with an agreed content under the responsibility of the different State Survey Agencies and the Institute for Applied Geodesy on the federal level, whereas the ALK databases are to be realized by the municipalities and counties.

ALK as a first generation G/LIS standard is based on hardware and software technology available at the beginning of the 1980s. ATKIS as a second generation G/LIS standard is based on the knowledge of the late 1980s. Therefore, further considerations will be based on ATKIS experiences (Brüggemann, 1992).

18.2 The ATKIS reference model

The philosophy of ATKIS as a cartographic production system is described in Figure 18.1. Because of unsolved automated generalization problems it is obviously best to describe the modelling of the landscape itself and its modelling according to a map in different data models and to store them in different databases. So the Digital Landscape Models (DLMs) contain the objects and phenomena on the earth's surface in their real place, even though simplified according to a defined accuracy. The Working Committee of the German State Surveying and Mapping Agencies (AdV) decided to build DLMs according to the map scales 1:25 000, 1:200 000 and 1:1 million and called DLM 25, DLM 200 and DLM 1000, respectively.

On the other hand Digital Cartographic Models (DKM) derived from the DLMs only contain those objects represented in the map to be realized. The limited space on the map makes it necessary to displace objects partly with the result that the position and shape of the object can be different from the real situation.

The definition phase was ended with the decision on the ATKIS standard by the AdV in 1989. Now we have:

- the ATKIS Object Class Catalogue for the DLM 25 and DLM 200,
- the specification of the DLM and DKM Data Models, and
- the ATKIS Symbol Catalogue for the DKM 25 and the map series 1:25 000 to be derived from the DKM 25.

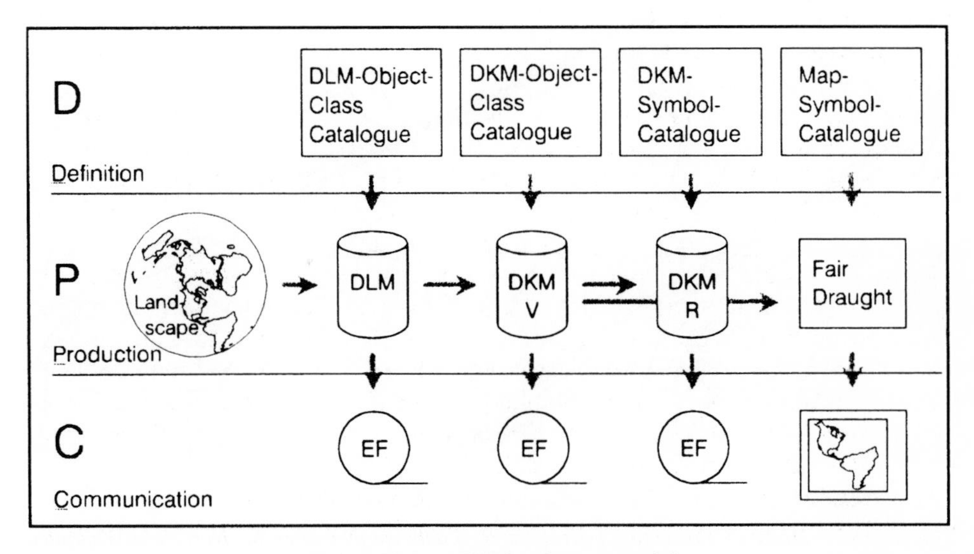

Figure 18.1 ATKIS reference model.

18.3 Strategies for cartographic products

The DKM is the classic cartographic product in the ATKIS philosophy. It can be understood as a digital, object-structured topographic map. In addition to the DKM, the Survey and Mapping Agency of North-Rhine-Westphalia plans to realize a map-oriented presentation of the DLM 25, which will be fully symbolized and generalized in a simple manner.

18.3.1 The DLM 25 Presentation Graphic (PG)

The PG is a map-oriented presentation of the DLM 25. Its symbolization is similar to the DKM symbolization. The PG will be generalized in a simple manner without any manual interaction. In this sense, it is planned to integrate algorithms for contour-simplification and area-combination directly into the presentation software. Here, generalization only changes the graphic representation of the data and not the information in the database. The Presentation Graphic is shown in Figure 18.2.

18.3.2 The Digital Cartographic Model (DKM)

The DKM only contains those objects represented in a topographic map. In contrast to the attribute-oriented object-structure of the DLM, the structure of the DKM is oriented to the graphic representation (Figure 18.3 shows the DKM object-parts of a forest) (Vickus, 1994).

However, a graphic-oriented DKM object-model, derived from the DLM alone does not fulfil the requirements of a topographic map. Symbolization and limited space on the map make it necessary to generalize the map features in a semantic, topologic and geometric sense.

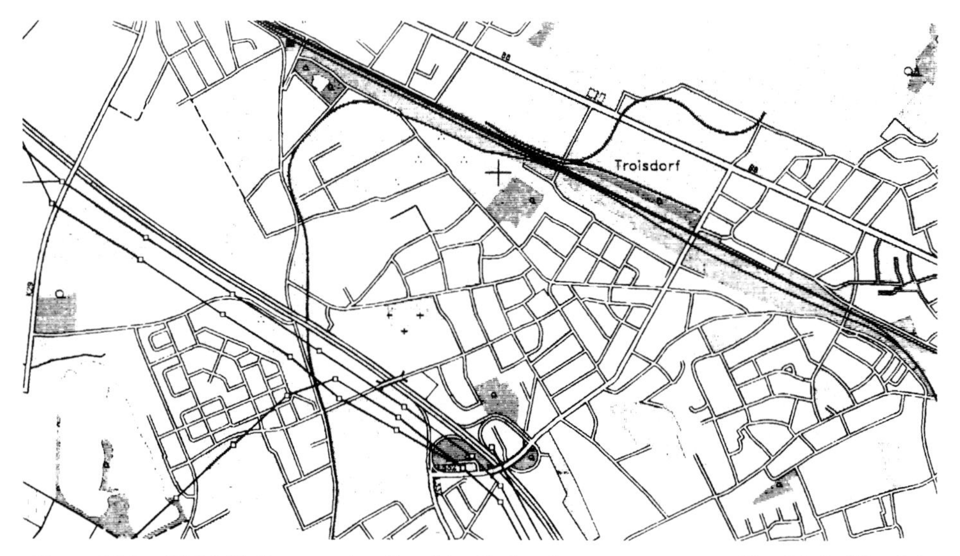

Figure 18.2 DLM 25 Presentation Graphic. © Landesvermessungsamt Nordrhein-Westfalen 1993.

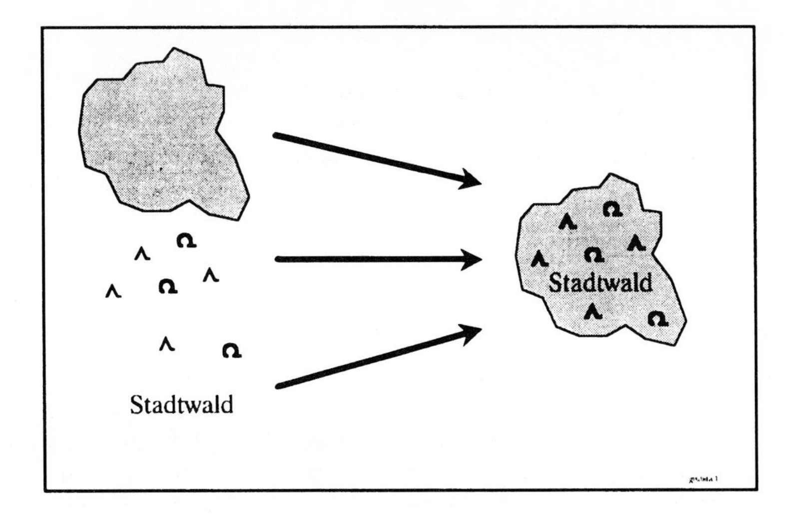

Figure 18.3 DKM object-parts.

The following considerations refer to a DKM 25 with areal representation of settlements. Single buildings in settlements are actually not contents of the DKM. In this sense, generalization includes:

- simplification and emphasizing of linear map features,
- simplification of area contours,
- classification and combination of area features,
- selection of point features,
- selection of linear features,
- selection of area features, and
- displacement.

Research and development for methods and algorithms to solve these generalization processes have not been completed until now. For simplification of linear map features the algorithm of Weber is preferred, the basic idea of which is a moving aperture (Weber, 1978). In order to select and combine areas, a method with geometric and semantic criteria is used. This method is described by Heisser *et al.* (1994). Other solutions are still being researched.

The results of generalization are:

- the change of geometry (e.g. by simplification, displacement),
- the change of topologic structures (e.g. by combination, selection), and
- reduction of the number of relevant map features (e.g. by combination, selection).

In contrast to the generalization of the PG as discussed above, the generalization of the DKM changes the information in the database. An example for the new, ATKIS-related topographic map is shown in Figure 18.4.

These two different cartographic products lead to the basic differentiation into two generalization levels. Low-level generalization will be a regular tool of a GIS, which generalizes during the presentation of the data. The generalization software will be available for different applications. High-level generalization software is special software for

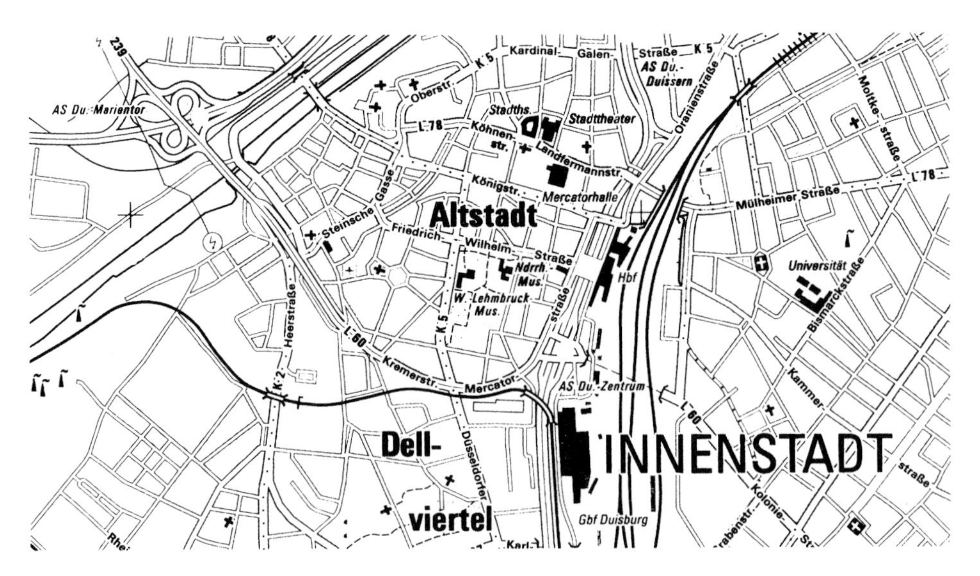

Figure 18.4 Graphic of the ATKIS-related DKM. © Landesvermessungsamt Nordrhein-Westfalen 1993.

special applications (e.g. for high-quality topographic maps). In this case, combination of interactive- and batch-processes is possible.

18.4 Development plans and research needs

The existing ATKIS solutions concentrate on the first DLM data-capturing phase, on the questions of how to provide data and how to produce an analogue representation of the DLM data content to interested users, without any manual operations. Based on the ATKIS implementation of North-Rhine-Westphalia on the ALK-GIAP, these questions have already been solved.

For ATKIS-related cartographic products, the next development steps will be:

1. **PG:** Improvement of the DLM Presentation Graphic (e.g. integration of low-level generalization software and rules for name placement).
2. **DKM-integration:** Integration of DKMs and processes supporting the transformation of a DLM to a DKM, until 1995, based on already-existing ALK-GIAP software components like its integrated object-related spatial data management facilities and especially its highly advanced special mapping language, which should be progressively developed towards a basic cartographic generalization language with an object-oriented approach.
3. **Undisturbed data flow:** Realization of an undisturbed data flow from DKM to the maps as analogue products based on the ALK-GIAP mapping language providing the tools for deriving map symbols according to the ATKIS Symbol Catalogue and based on software and hardware products supporting the digital pre-printing process available on the market. This problem will require solving until 1995, too.

The previous chapters showed that the ATKIS production process shown in Figure 18.1 can be realized until 1995 without further research activities. Further research needs arise from economic reasons in combination with such quality aspects as accuracy, reliability and up-to-dateness of the ATKIS data and their vertical and horizontal integration as described. The production process planned for 1995 needs a lot of interactive support by operators.

Scientific research has to give answers to problems which cannot be solved because of a lack of basic knowledge within a science. From the ATKIS point of view this seems to be valid for the following fields:

1. **Data content definition:** The main requirements related to data content definitions are seen in the lack of semantic modelling possibilities. What is missing is the transfer of knowledge from object-oriented approaches and artificial intelligence developments of computer science to cartography. This adaption could be done very soon because of the availability of these techniques. More complicated but directly related to cartographic science is the fundamental analysis of semantic descriptions of earth-related objects and phenomena and their integration in the DLM and DKM concepts as specific cartographic experts' knowledge. This knowledge is urgently needed for highly advanced production processes like digital image analysis, comparison of a changed image with an existing DLM, and derivation of a DKM from a DLM by automated cartographic generalization.

   Computer science is asked to develop a formal data description language following the above-mentioned requirements of semantic modelling.
2. **Data exploitation:** From the author's point of view, deriving digital and analogue products from DLM and DKM data has to be a main research field for the next decade. There has been done a lot of research in the field of cartographic generalization during the last decade which this author feels remained unsatisfactory because of their restriction to partial solutions. The problem seems to be that cartographers need a new kind of thinking that takes into account the basic differentiation into DLM, DKM and graphical products. Map measurement fades behind map layout because of the new possibilities offered by quantitative DLM analysis. Cartographic knowledge on generalization has to be collected and formalized by a formal language based on ideas of object-oriented thinking and artificial intelligence.

   On the other hand, cartographers should concentrate on research in the field of map layout, too. The new digital techniques allow more flexibility of map production techniques and a large variety of map products using new knowledge of cartographic symbol theory.

18.5 Outlook

Map production is an important, but not the only, goal of a topographic–cartographic information system of the Surveying and Mapping Agencies, considering the upcoming use of modern digital techniques by the traditional users of our data, too. Already the demand for basic digital information on the earth's surface has become so large that there are problems satisfying all needs. With great efforts, traditional users of the basic data, who until now used maps, can build their own application-oriented information systems like road and traffic information systems, planning information systems, environment protection information systems, and soil information systems. They all rely on the basic data provided by surveying and mapping. In this sense, it is important to carry out generalization not only for production of topographic maps. In future, generalization should be a variable tool of Geoinformation Systems.

References

AdV, 1989, Amtliches Topographisch–Kartographisches Informationssystem (ATKIS). Das Vorhaben der Landesvermessungsverwaltungen zum Aufbau Digitaler Landschaftsmodelle und Digitaler Kartenmodelle, Bonn: Arbeitsgemeinschaft der Vermessungsverwaltungen der Bundesrepublik Deutschland (AdV).

Brüggemann, H., 1992, Germany's cartographers recognize the need for digital data research and cooperation, *GIS Europe*, **1**(6), 34–7.

Grimm, W., 1993, Neue Kartengraphik für ATKIS DKM 25, *Kartographische Nachrichten*, No. 2, pp. 61–8.

Heisser, M., Vickus, G. and Schoppmeyer, J., 1994, The generalization process selection and combination of small areas — a rule-oriented arrangement, Paper to the *GISDATA-Specialist Meeting on Generalization*, Compiegne, France; also published in this book — Chapter 11.

Jäger, E., 1990, 'Untersuchungen zur Kartographischen Symbolisierung und Verdrängung im Rasterformat', Wissenschaftliche Arbeiten der Fachrichtung Vermessungswesen der Universität Hannover, No. 167, Hannover.

Powitz, B.M., 1993, 'Zur Automatisierung der Kartographischen Generalisierung topographischer Daten in Geo-Informationssystemen', Wissenschaftliche Arbeiten der Fachrichtung Vermessungswesen der Universität Hannover, No. 185, Hannover.

Vickus, G., 1994, 'Digitale Topographische und Kartographische Modelle sowie Entwicklung ihrer Überführungsstrukturen am Beispiel von ATKIS', *Schriftreihe des Instituts für Kartographie und Topographie der Universität*, Bonn.

Weber, W., 1978, Liniengeneralisierung und Datenreduktion unter dem Gesichtswinkel der mathematischen Optimierung, *Nachrichten aus dem Karten- und Vermessungswesen*, pp. 55–66, Reihe I, No. 74, Frankfurt.

Index

a posteriori evaluation 65
a priori evaluation 65
active world 85
ad hoc evaluation 65
adaptive learning 141
ALK-GIAP 250
amplified intelligence 9, 63, 139, 170
  extensions 142–5
analysis of graphical documents 62
analysis of text documents 62
arc tree 125
area partitioning 120, 121
  hierarchy 121
areal features 153
artificial intelligence 10, 135–47, 137, 144
  drawbacks 138
artificial vision 79
ATKIS 25, 49, 54, 148, 149, 151–3, 246–52
  data exploitation 251
  development plans and research needs 250–1
  reference model 247
  strategies for cartographic products 248–50
attribute accuracy
  CALVEG map 198–9, 202–5
  in natural resource maps 194–206
  transect analysis of 198–9
attribute transformations 214
attributes 93
automated generalization 4, 5, 7, 8, 11–15, 56–69, 140, 184
  current attemps 237
  data quality 190–1
  early attempts 237
  need for 238–9
  paper versus VDU 238
  scale-free or fixed-scale series 238
  update 238–9

batch generalization 8–9, 211
benchmarking 65, 127
Binary Line Generalization tree *see* BLG-tree
Blackboard Systems 140
BLG-tree 120, 124–7, 129, 130

CALL-frequency 54
CALL-sequences 54
CALVEG map 197–9
  attribute accuracy 198–9, 202–5
  classification error 199
  classification system 197
cartographic expressions 48
cartographic model theory 47–55
  primary models 47
  secondary models 47
  tertiary models 47
cartography
  automated 24
  traditional map-centered 29
  traditional process 22–4
case-based reasoning (CBR) 135, 137, 142, 145
CHANGE program 8, 51, 54, 238
characteristic points 78
cluster analysis 168, 220
clustering 130
cognitive engineers 138
cognitive task analysis 138
communication relations 83
communication technologies 47-8
complex objects 76
computer-assisted generalization 47–55, 219
connexity relations 79–80
contexts 114–16
controlled data reduction 57

data abstraction 6
data aggregation 43–4
data generalization, modelling 87
data models 13, 15, 92
  scale-specific 95
data quality 183, 184, 186
  and generalization 14
  automated generalization 190–1

data quality (*cont'd*)
  in generalization research 12
  model generalization 188–9
data reduction 183
data structure 6, 13, 15, 120
  for areal features 153
  OSGEN 3 241
data transformations 183
databases 24–9, 56, 92, 185
  and map display 24
  compilation 26
  defining 26
  derivation 57
  information content 25
  multiple representation 210–11
  source to geographic 25
  structured initial 45–6
  to database transformations 27–8
  to map transformation 28–9
declarative knowledge 51, 54
declarative model 151, 152
deferred mode 242–3
Delaunay triangulation 106
developed areas workflow 233
Digital Cartographic Models (DCM/DKM) 48, 53, 57, 148, 152, 247–9, 251
Digital geodata
  generalization for presentation on screen 38–40
  generalization for production of paper maps 32–8
Digital Intermediate Models (DZM) 148
Digital Landscape Models (DLM) 148, 247
Digital Line Graph (DLG-E) data model 25, 93–102
  case studies 96–100
  scale-dependent extensions 95–6
digital object model (DOM) 48, 49, 52, 53
digital object-oriented landscape model (DLM) 48, 49, 53, 57, 152, 251
Digital Orthophotoquad (DOQ) 96
digital terrain models (DTMs) 57
displacement 111–12
display 183, 184
  quality 184
distance–direction matrix 164–6
distortion 66
distribution structures 82
  modelling 81–2
'Divide and conquer', OSGEN 3 241
DLM 25 presentation graphic (PG) 248–9
DLMS DFAD 126, 127, 129
domain expert 162
Douglas–Peucker algorithm 125, 188, 190

entity 93
evaluation 136
expert system shell 10, 85
expert systems 45, 135, 137, 162

Fairview line 167
feature categories 153
feature class dependency 221
feature dimension dependency 221
features 94, 242
fractal analysis 78
frame-based representation 107
free-space triangles 109
functional model 151, 154–5

GAP-tree 120–32
  building 121–3
  operations 123–4
generalization 6–8, 10, 12, 50, 51, 183, 184
  academic research 13
  and data quality 12, 14
  and knowledge-based approaches 10–12
  context of 21–2
  control 117
  definition 236
  directions for future research 14
  effect on attribute accuracy in natural resource maps 194–206
  effects of 184–8
  essence and concerns of 220–1
  evaluation of alternatives 64-7
  evaluation of process 136–7
  events and rules 11
  experimental project 222–34
  facilities provided 4
  first stage 82
  for and against 3–5
  for presentation on screen 38–40
  for production of paper maps 32–8
  formalization 10, 220–1
  graphics-oriented 7
  hierarchical top-down bottom-up approach 235–45
  importance of external information 44
  in GIS 31–46, 50–4
  interpretation of human process 220
  intuitive components 65
  issues concerning 5
  knowledge needed for 74–6
  line 8
  management 117

generalization (*cont'd*)
  map images composed of several components 40–3
  map-oriented 23
  model-oriented 6–7, 12, 43–5
  modelling 47, 83–5
  modules 46, 214–16
  objectives and characteristics 13
  operators 14, 82, 88, 214
  potential workflow 222
  present and future development 12–15
  problems 13, 27
  process 183
  purposes and benefits of 31
  quantifying effects of 183–93
  research and development 52
  second stage 82
  semi-automated 4
  state of the art 3–17
  survey 83
  systematic overview 50
  unpredictable effects 12
generalization index 6
Generalized Area Partitioning (GAP)-tree *see* GAP-tree
generic knowledge 75
generic modelling 79
genetic algorithms 136, 144
  evaluation of 144–5
GE0++ 126, 127
geographic dataset
  to geographic dataset 25
  to map 25
Geographic Infomation Systems *see* GIS
geometric distribution 81
geometric knowledge
  from encoding process 164
  from existing rule bases 162–3
geometric layer 84
geometric modelling 77
geometric properties 166–7
geometric structures 81
geometrical measures 66
geomorphic properties 166–7
GIS
  applications 56
  cartography based on 24–9
  concept of 48–50
  data visualization 3
  generalization in 50–4
  implementation 48, 49–50
  set-up 49
global characterization 78
global measures 66
graphical documents, analysis of 62
graphical representation 92

holistic generalization 106–19
holistic solutions 136
Honylake line 167
human–computer interaction 14–15, 135, 138, 219
hybrid systems 140
hydrography data generalization workflow 222–33
Hypergraph Database Structures (HDBS) 11

IF–THEN rules 141
implicit knowledge 140
incremental generalization 211–14
  ordering steps 215
  principle of 211
  updating 213–14
index structures, average response time 129, 130
individual buildings workflow 233
inferencing procedures 141
initialization information 153
Institute of Cartography (IC) 54
interaction logging experiments 64
interactive generalization 9–10, 45–6, 63, 219
interdisciplinary expert knowledge 150
INTERGRAPH 8, 10
interviewing 162

kappa 202–3
  environment 108
  inheritance mechanism 117
knowledge acquisition 56, 58–64, 135, 161–79
  methods 60–1
knowledge-based modules 45–6
knowledge-based simplification 77
knowledge-based software methods 151
knowledge-based systems 10–12, 58–9, 135, 137–41, 149–50
  second generation 139
  short-comings 137–9
knowledge compilation 140
knowledge elicitation 140–2, 236
knowledge engineer 141, 161–2
knowledge engineering 59, 62, 135, 136
knowledge formalization 15
knowledge implementation 135
knowledge needs in generalization 150–1
knowledge representation 135, 139–41
knowledge types 162–8
knowledge used by cartographer 75–6

Landsat Multi-spectral Scanner (MSS) 197
large-scale maps 239
layer, definition 84

learning by being told 162
learning by observation 162
line, global description 77–8
line filtering algorithm 58
line types identification 79
linear object 77
Lisp 10

machine learning 62–3
MAGE system 106–7, 114, 117, 118
Maïda2D 85
manual generalization 220–1, 236
map compilation process 21–2
Map Generalization System (MGS) 171, 175
map series, change between scales 11
mapping rules 28
merging of objects 112–13
meta-objects 241–2, 242
MGE Map Generalizer (MGMG) 10, 219, 220
minimal dimensions 66
Minimum Mapping Unit (MMU) 194, 200
model generalization 51, 56–8, 184
  data quality 188–9
  research and development 51
modelling
  data generalization 87
  for generalization 47, 83–5
Modifiable Areal Unit Problem 92
modus ponens 137
multiple representation 210–11
  database 210–11
multi-scale line tree 125

National High Altitude Photography series (NHAP) 201
national mapping agencies (NMAs) 4, 11, 53, 73, 92, 163
natural resource maps, attribute accuracy in 194–206
neural nets 63, 136, 137, 140, 143
  evaluation of 144–5

object entries 125
object generalization 50
object-oriented map generalization 91–105
object-oriented programming 6
object properties and relations 88
objective generalization 136
objects 75, 84, 93, 241–2
  geometric distribution 80, 81
  geometric properties 75–9
  merging 112–13
  number of 83
  scaling 110
  simple and complex 83
  spatial distribution 80
OEEPE 32
on-the-fly maps 120–32, 238
operational logic 221
operators 136
orders 215
Ordnance Survey 186
OSGB
  generalization rules 240
  large-scale maps 239
OSGEN 1 240
OSGEN 2 240
OSGEN 3 240–3
  data structure 241
  'divide and conquer' 241
  preliminary results 244

paper maps 32–7, 238
polynomial function 79
Postgres DBMS 124, 125
Postgres query optimizer 127
Postquel queries 126, 127
preprocessing operation 57
problem definition 149–50, 152
problem-solving strategy 158
procedural knowledge 51, 53, 54, 141, 150, 168, 171
procedural system 51
process efficiency 221
processes 136
prolog 10
propose-test-and-adapt method 145
proximity relations 80
  modelling 81

Quad-tree indexing technique 85
qualitative evaluation 67
quality assurance 65
quantitative measures 66

reactive data structures 120, 124–6
Reactive-tree 120, 124, 126, 127, 129, 130
recognize–understand–translate process 114
recognizers 114–17
referencing data 40–3
relational tables 99
relationships 94
representational redescription 140
rule-based system 74–5, 162–3
rule-orientated definition of small area 'selection' and 'combination' 151–8
rule-orientated model 151, 155–8

SA/SD method 211
sampling frequency 200–3

scale effects 46, 74, 95–6, 163, 186, 187, 189, 200
scale factor 23–4
screen presentation 238
  comparison of vector and raster data scaled down 40
  feature selection and symbolization 43
  minimal dimensions of data displayed 39–40
  overall image 40
  zooming out for location 38–9
semantic knowledge 150
semantic network 11
semantic relations 82–3
  modelling 83
  types of 83
shapes, description and characterization 78
simplices 108
simplicial complexes 108
simplicial data structure (SDS) 106–10
simplicial data structures (SDS)
  displacement 111–12
  entity–relationship diagram 109
  implementation 108–9, 117
  major operators 110–13
skeletonization 111
smoothing 221
  using B-spline 172
  using five-point weighted moving average 172
societal mapping 47, 52
software-related measures 66–7
soil generalization process 28
spatial accuracy 221
spatial analysis techniques 27
spatial arrangement relations 80, 81
spatial buffering 92
spatial conflicts
  definition 109
  detection 109
  resolution 109–10
Spatial Data Transfer Standard (SDTS) 93, 188
spatial indexes 81
spatial relations 79–82, 241–2
spatial resolution 87
spatial transformations 214
specifications 65
street and house generalization 23
strip tree 125
structural constraints 79
structural knowledge 15, 150
structural layer 84
structure recognition 15
structure signatures 63
surface generalization 92
Swiss National Atlas 32
system orders 155

tacit knowledge 138, 140
teachback method 141, 145
temporal abstraction 7
test-and-adapt method 145
text documents, analysis of 60
topologic layer 84
topologic maps 80
topologic objects 80, 84
topological measures 66
transect analysis of attribute accuracy 198–9
tree entries 125

urban area 23
user-interface design 168–78

VDU *see* screen presentation
vector of displacement 110
vegetation data generalization workflow 222
vegetation maps 195–7
  fundamental properties 200
vegetation properties 195–7

work psychology 135
world of reference 85

zoom function 172